Faded Splendour, Golden Past

Faded Splendour, Golden Past

Urban Images of Burma

ELLEN CORWIN CANGI

KUALA LUMPUR
OXFORD UNIVERSITY PRESS
OXFORD SINGAPORE NEW YORK
1997

Oxford University Press
Oxford New York
Athens Auckland Bangkok Bombay
Calcutta Cape Town Dar es Salaam Delhi
Florence Hong Kong Istanbul Karachi
Madras Madrid Melbourne Mexico City
Nairobi Paris Shah Alam Singapore
Taipei Tokyo Toronto

and associated companies in
Berlin Ibadan

Oxford is a trade mark of Oxford University Press

Published in the United States
by Oxford University Press, New York

First published 1997

British Library Cataloguing in Publication Data
Data available

Library of Congress Cataloging in Publication Data
Cangi, Ellen Corwin.
Faded splendour, golden past: urban images of Burma/Ellen Corwin Cangi.
p. cm.—(Oxford in Asia paperbacks)
Includes bibliographical references and index.
ISBN 983 56 0010 4 (paper)
1. Pagan (Burma)—History. 2. Mandalay (Burma)—History.
3. Rangoon (Burma)—History. I. Series.
DS530.9.P33C35 1997
959.1—dc20
96-30596
CIP

Typeset by Indah Photosetting Centre Sdn. Bhd., Malaysia
Printed by KHL Printing Co. (S) Pte. Ltd., Singapore
Published by Penerbit Fajar Bakti Sdn. Bhd. (008974-T),
under licence from Oxford University Press,
4 Jalan U1/15, Seksyen U1, 40000 Shah Alam,
Selangor Darul Ehsan, Malaysia

Acknowledgements

THINKING about cities, particularly cities in Burma, has occupied a good deal of my time during the past several years. As I researched and wrote this book, I frequently recalled my days in graduate school at the University of Cincinnati. Those were wonderful years full of personal growth, friendship, and discovery. I will always be indebted to my teachers, Zane L. Miller and Henry D. Shapiro. Thanks also to Judith Spraul-Schmidt, Robert Fairbanks, and Alan I. Marcus, who shared those terrific years with me.

More recently, a number of people here in Singapore helped bring this project to fruition. The late Michael Sweet was truly a remarkable man and scholar, always ready to search the world over for some long out-of-print source. Fellow researchers from the Friends of the Museum, Sally Ward and Maya Jayapal, cheerfully supplied a much needed but gentle nudge now and again. In the final stages of preparation, Julie Yeo of Antiques of the Orient generously offered her expertise in preparing illustrations.

I am also indebted to my family. Fortunately, they share my love of cities. Together we have done London, Bangkok, Cairo, Hong Kong, Boston—the list goes on and on. They are—all three of them—good company and for that I am endlessly grateful.

Singapore
June 1996

ELLEN CORWIN CANGI

Contents

1
Introduction

Burma's History: The Long View

AT first it was only a trickle, small groups of tribesmen slowly making their way into Burma across the rugged mountains in the north and down the broad river valley to the dry central plateau. What caused the Burmans to leave their ancestral home in East Asia somewhere between the Gobi Desert and north-eastern Tibet no one knows. By the ninth century, however, the trickle had turned into a torrent until they finally outnumbered other ethnic groups. The Burmans were, in fact, relative latecomers to the land which they would one day claim as their own. They settled first on the Kyaukse Plain and from there spread out, west across the Irrawaddy River, north up the Chindwin River valley, and south towards Prome to infiltrate and eventually conquer Burma (Figure 1).[1]

The Mranma, as the tribes called themselves, quickly encountered resistance from the Mons, a people whose hostility to the Burmans forms a strong central theme in the country's history. The Mons, close relatives of the Khmers, dominated the southern delta of Burma and had yet to be challenged.[2] They had arrived in Burma some two hundred years earlier from the kingdom of Dvaravati in neighbouring Siam. Culturally more advanced than the Burmans, they had established a capital at the port city of Thaton and maintained a vast trading network with contacts as far away as India. The Mons were an innovative people whose agricultural techniques formed the economic backbone of later Burmese empires. They constructed a sophisticated irrigation system on the dry Kyauske Plain, planted a variety of crops, including cotton, maize, and millet, and introduced wet-rice cultivation. In addition, they planted beans, an abundant and basic source of protein. Consequently, it became possible to sustain population growth on the Kyaukse Plain which became the centre of imperial Burma.[3]

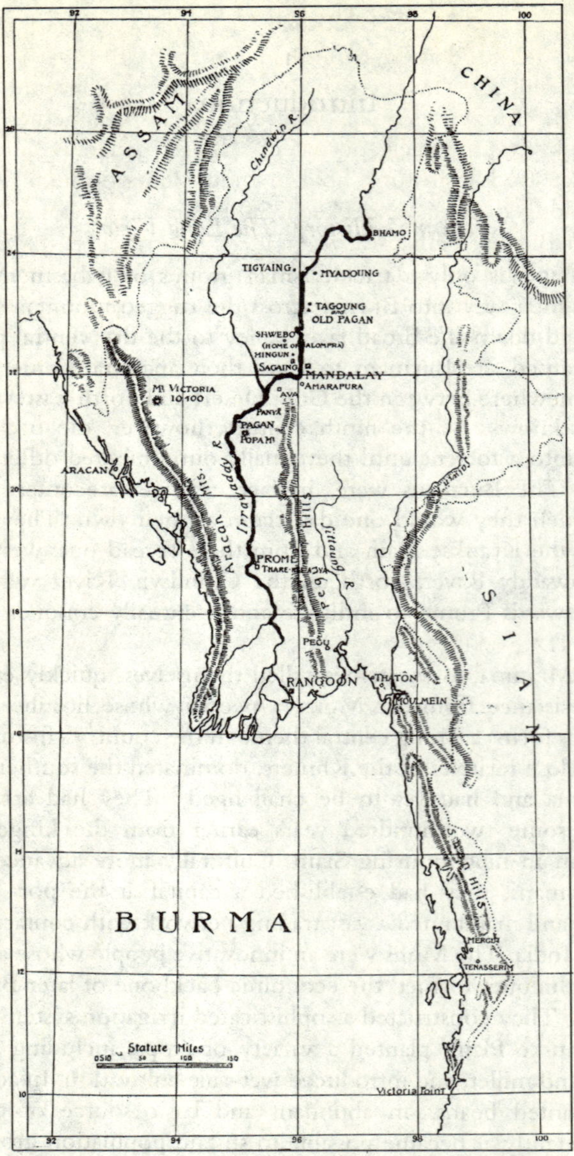

Fig. 1 Sketch map of Burma. (O'Connor, *Mandalay and Other Cities*, 1907)

Legend tells us that the Mons had contacts with India as early as the third century BC when two Buddhist monks, Sona and Uttara, sent by the Emperor Asoka, arrived in Thaton. Whether or not the monks' visit actually occurred, scholars generally agree that Theravada Buddhism took hold in Indo-China through repeated and prolonged contacts between South India, Ceylon, and the Mons of Burma. Perhaps it was traders and travellers, rather than monks, who had such a great impact on Mon culture; at any rate, the adoption of the Pali script, kingly rituals, temple architecture, and sculpture all bore the unmistakable marks of Hindu influence.[4]

In contrast to the Mons, with their clear cultural identity, were the Pyus, who merged with the Burmans after a brief appearance in history. Some scholars even suggest that the Pyus might have been the 'vanguard of the Burmans,' so closely do the two groups appear to have been related.[5]

As early as the seventh century, the Pyus established a kingdom at Sri Ksetra in southern Burma. Ruins of the city's walls indicate that it was even larger than Pagan or Mandalay. Temple ruins featuring vaulted ceilings reveal an advanced civilization. The Pyus' religion was a pragmatic and creative blend of old and new; Buddhism, Hinduism, *naga* worship, and animism existed side by side. Socially, the Pyus were a loosely knit group of tribes and, unlike the Mons, had neither hereditary chiefs nor a nobility, factors which may help to explain their demise. According to Chinese sources, Sri Ksetra was plundered in the year 832 and 3,000 Pyu captives were sent in bondage to Yunnan in China. The survivors are thought to have merged with the Burmans. At any rate that is the last we hear of the Pyus.[6]

According to *The Glass Palace Chronicle of the Kings of Burma*, compiled in 1829, Burma traces its royal lineage to the house of the Buddha Gautama. Forty kings, identified only by name, ruled before the eleventh century. Burma's national history actually began with the reign of King Anawrahta (r. 1044–77), founder of the First Burmese Empire. King Anawrahta's huge empire, with its capital at Pagan, foreshadowed modern Burma both in size and shape and lasted for well over two centuries.[7] Anawrahta also introduced the Burmans to Theravada Buddhism, the religion of the Mons. In comparison to the more complex Mahayana Buddhism, the Theravada Buddhists stressed asceticism, meditation, and the monastic life. King Anawrahta hoped to reform the dissolute but powerful Buddhist clergy at court with this simpler form of Buddhism.[8] The Mons refused to share their religion and a full-scale invasion of Thaton was launched in order to obtain the Theravada scriptures.

In addition to the Mons' religion, King Anawrahta and the Burmans found many other aspects of their culture appealing. For their part, the Mons became Pagan's most-hated adversary, ever ready to launch a rebellion. Their culture, however, was irresistible. The royal court at Pagan adopted Pali as the official language as well as the Mon alphabet. Moreover, many of the pagodas built on the vast plains surrounding Pagan had a distinctively Mon style. Bolstered by a strong infusion of the more advanced Mon culture, Pagan's golden age had begun.[9]

King Anawrahta's immediate successors in the Pagan Dynasty were Kings Sawlu, Kyanzittha, and Alaungsithu, names which evoke awe-inspiring images of oriental courts where ritual, ruthlessness, and intrigue combined to ensure royal power. Generally, the crown passed from father to son with one significant exception, Kyanzittha (r. 1084–1112), who turned out to be one of the dynasty's most capable kings. In a strange twist of fate, Kyanzittha, King Anawrahta's trusted general for many years but who had fallen from grace, succeeded to the throne after King Sawlu (r. 1077–84), the legitimate heir, died at the hands of Mon rebels. At his coronation, King Kyanzittha went to great lengths to establish himself as the legitimate ruler of Pagan. Inscriptions indicate that he consciously sought to unify the empire which he had inherited from King Anawrahta.[10]

When King Kyanzittha died in 1112, he was succeeded by his grandson, King Alaungsithu (r. 1112–67), who reigned for over fifty years. Although the king himself was half Mon, there was a marked decline in Mon influence at court as Burmese culture flourished. Burmese supplanted Pali as the court language and was used for the first time in literary works. Similarly, temple architecture took on an indigenous character. Most significant of all was the fact that Theravada Buddhism became the religion of the common people, exerting a strong moral influence on their daily lives. Monks from a new monastery at Nyaung-u set out to convert the people of Burma and neighbouring countries. Consequently, when Islam later spread to Malay and Indonesia, it altogether failed to appeal to the Buddhists of Indo-China.[11]

The Pagan Dynasty produced a long list of capable rulers who, perhaps unwittingly, undermined the empire's future by transferring large amounts of wealth to the clergy for the construction and maintenance of religious monuments. The thirteenth century brought the final disaster when the army of the great Kublai Khan invaded Burma in 1283 after the short-sighted king haughtily refused demands to pay tribute. After that, control of Burma was divided among the Mons in the south, the Shans on the Central Plateau, and the Arakanese in the west for nearly three centuries.[12]

The Toungoo Dynasty and the Second Burmese Empire

In the wake of Pagan's destruction, Shan tribesmen poured out of the hills into the Irrawaddy Valley to defeat the Mongols and subsequently fill the power vacuum in central Burma. By the turn of the fourteenth century, the so-called Three Shan Brothers (actually three factions, each of which controlled a different geographical area) had defeated the Mongols and ruled Upper Burma. Arakan regained its independence, and the Mons controlled Lower Burma from their capital at Pegu. Expansionist Ayut'ia (Siam) threatened the Mons, and, periodically, the Chinese intervened to check the strength of the Shan. Burma was anything but politically stable.

Late in the fifteenth century, King Minkyino (r. 1486–1531) and the Burmans began building another power base at Toungoo in the Sittang Valley. The Shans, intent on tearing central Burma apart, took little notice of the new kingdom until it was too late. After King Minkyino's death, his brilliant son Tabinshweti (r. 1531–50), rescued Upper Burma, conquered the South, and founded the Second Burmese Empire. King Tabinshweti, unlike earlier kings who ruled from the centre of the country, established his court in the south of Burma at the city of Pegu.[13]

Murdered at the age of thirty-two, King Tabinshweti was succeeded by his trusted general and brother-in-law, Bayinnaung

(r. 1551–81), a king greatly admired by rulers of the later Konbaung Dynasty. He not only successfully united the Burmans, Shans, and Mons under one crown but also conquered Ayut'ia and Chiang Mai in neighbouring Siam, extending the empire's boundaries to their furthest limits. Unfortunately, the empire was founded solely on military might and fell apart when King Bayinnaung died.[14]

Once again Burma descended into chaos as civil war visited the country. Attracted by the power vacuum, Europeans began arriving in large numbers in the seventeenth century. In turn, the Portuguese, Dutch, British, and French set up trading posts along Burma's coastline. While most came as traders, others, like the legendary Philip De Brito y Nicote, entertained fancies of becoming king in this strange and exotic land. De Brito began his career as a cabin boy on a Portuguese ship. Later, having convinced the King of Arakan of his not inconsiderable financial talents, he became customs collector in the port city of Syriam. With his new-found source of funds, the treacherous De Brito fortified the town and proclaimed himself king. His reign lasted for thirteen years. Then in 1613, King Anaukpetlun (r. 1606–28), of the Toungoo line, mounted an attack on Syriam. The siege lasted thirty-four days before De Brito was finally captured.[15]

King Anaukpetlun was murdured the following year by his brother Thalun (r. 1629–48) who managed, but only at great cost, to restore peace and order to the ravaged land. In the wake of yet another failed Mon uprising, the new king drove thousands of Mon refugees into Siam. More important in the long run, however, is the fact that King Thalun transferred the kingdom's capital from the port city of Pegu to Ava in the interior of the north. Neither he nor his successors had much use for the outside world. The Burmese kingdom became permanently centred in Upper Burma and, increasingly, isolationism, xenophobia, and traditionalism characterized court policies. Trade was left in the hands of foreigners, a fact which would later have serious ramifications for the nation. To make matters worse, during the middle decades of the seventeenth century, Burma was ruled by a series of weak kings culminating in the demise of the Toungoo Dynasty and the Second Burmese Empire in 1752.[16]

The Konbaung Dynasty and the Third Burmese Empire

Soon, however, a village chief from central Burma, who took the name of Alaungpaya (Alompra) (r. 1752–60), established what became the third and last Burmese Empire under the Konbaung Dynasty. King Alaungpaya's rise to power was meteoric. In December 1753, he conquered a combined force of Mons and Shans at Ava and the following May, he reached Prome. Having taken Arakan, he accepted the allegiance of the Mon leaders of Henzada, Myaungmya, and Bassein and then turned south to finish off what was left of the ever-recalcitrant Mons. A year later, King Alaungpaya decisively defeated them at Dagon, and, to commemorate the auspicious occasion, founded Rangoon.[17]

Unfortunately, peace eluded Burma. Following victory against Pegu in the spring of 1756, King Alaungpaya could not resist a final campaign against Siam, despite the fact that royal astrologers warned of disaster. His troops surrounded Ayut'ia in April 1760 and victory appeared to be at hand. But fate intervened and the king was critically wounded. Upon hearing the news, the army retreated in great confusion. King Alaungpaya died at the age of forty-six, leaving the Burmans with 'a taste for military glory which for over a half century was to make them the terror of their neighbours'.[18]

King Alaungpaya's successor, his younger son Hsinbyushin (r. 1763–76), moved the royal court from Shwebo to the old Shan capital of Ava and soon undertook another campaign against Ayut'ia. In the spring of 1767, the Burmans plundered and burned the Thais' capital city, and stood poised to take the country. The Chinese intervened on behalf of the Thais at the last minute, however, and returned a balance of power to the region. Stubbornly, King Hsinbyushin refused to admit defeat. War dragged on until his death in 1776, resulting in nothing but hardship and misery for the people of Burma.[19]

King Alaungpaya's third son, Bodawpaya (r. 1782–1819), won the throne but only after a bloody power struggle in which he massacred his rivals, including his own brothers and their entire families. A brother who managed to escape was later captured and put to death along with every single one of his supporters, their families, and servants. The village where the conspirators met was totally decimated—people, animals, crops, trees—everything was destroyed as a grim reminder to those who might be tempted to plot treason in the future. 'Hundreds of both sexes, and even some Buddhist monks, on vague suspicion that they had been privy to the conspiracy, were burnt alive upon an immense pile of wood.'[20] King Bodawpaya, the first of the Konbaungs to evince symptoms of insanity, thus inaugurated a savage tradition emulated by all of his successors except one.

To his subjects, King Bodawpaya, whose ministers compiled Burma's first census, was known as the 'Lord of the Great Law'. In contrast, the British diplomat Michael Symes viewed him as 'a child in his ideas, a tyrant in his principles, and a madman in his actions'. The king collected and piously studied Sanskrit texts, but at the same time severely punished his people for the merest infraction of Buddhist tenets.[21]

In 1784, King Bodawpaya attacked and annexed Arakan, a strategic province bordering British India. Then he conscripted Arakanese labourers by the thousands, subjecting them to forced marches across Burma's rugged terrain for the sole purpose of building pagodas at the capital Ava. Growing numbers of refugees crossed the border into British India and hostility between the two countries increased. Relations with the British worsened when King Bodawpaya refused to receive the envoy Michael Symes for several months. The border area became a source of almost constant conflict. The situation deteriorated even further when King Bodawpaya's ineffective grandson, Bagyidaw, succeeded to the throne in 1819. Full-scale war finally broke out on 4 March 1824. Despite heavy losses, mostly from disease and poor logistics, the British won what later became known as the First Anglo-Burmese War. In the Treaty of Yandabo, signed in February 1826, Burma was forced to cede both Arakan and Tenasserim plus critical areas of Assam and Manipur along the border with India. An unbelievable indemnity equal to a million pounds sterling was also imposed on the Burmans.[22]

Burma was left battered and bruised but by no means totally crushed. Unfortunately, the king lost himself in fits of melancholia until he was judged unfit to govern. In 1837, his brother

Tharrawaddy seized the throne and, following in his brother's bloody footsteps, ordered the massacre of his rivals, a practice which was rapidly depleting the ranks of competent advisers and ministers. To no one's surprise, King Tharrawaddy refused to recognize the Treaty of Yandabo and was reported to be amassing supplies and preparing his army for war. Rumours also reached Calcutta of the new king's bouts of deep depression, his uncontrollable temper, and, worst of all, his extreme cruelty.[23]

In 1845, King Tharrawaddy, now violently insane, was confined to the royal palace, and another struggle for the throne began. The winner, Pagan Min, like his father and uncle before him, ordered the deaths of his brothers and all members of their households. Moreover, he was also cruel like his father and uncle. King Pagan Min reportedly enjoyed shooting, using his subjects as moving targets. Once, to the horror of his court, he seized a bystander, pulled out his dagger, and carved a chessboard on the man's back. At his bidding, the king's ministers ordered the deaths of nearly 6,000 Burmans on false charges in order to confiscate their land. Anxious British merchants repeatedly appealed to their government in India to protect their holdings. It was only a matter of time before the British would once again intervene in Burma's affairs.[24]

In the summer of 1851, two British captains, whose ships were moored at Rangoon, were charged with embezzlement and murder. The charges were trumped up and everyone went free once bribes had been paid. This time, however, the Burman Govenor of Rangoon had gone too far in humiliating British subjects. Lord Dalhousie, Govenor-General of India, sent three warships, commanded by Commodore Lambert, up the Rangoon River to demand compensation and the removal from office of the govenor. King Pagan Min's ministers hastily agreed to both demands. Unfortunately, tempers were too hot and the matter did not end there. One thing led to another, insulting remarks were exchanged, and Lambert blockaded the port. Shots were fired from the shore, and the now infuriated Lambert ordered a full blast of artillery fire in retaliation. After he systematically destroyed every Burmese warship on the river, he set sail for Calcutta.[25]

To be sure, Lambert had overstepped his orders and he was duly reprimanded. But the damage was done; war was unavoidable. 'We can't afford to be shown to the door anywhere in the East,' Lord Dalhousie wrote, once again revealing the thin skin of the British. An ultimatum, demanding heavy compensation, was ignored by King Pagan Min and the Second Anglo-Burmese War began in April 1852. This time the British were better prepared to supply troops. (In fact, Britain's mortality in the second war against Burma was lower than peacetime mortality in India.) British troops occupied the southern ports of Rangoon, Martaban, and Bassein before the monsoon season began in May. The fate of the rich delta in the south, Lower Burma, was settled when the British annexed the province of Pegu on 20 December 1852, presenting it as a *fait accompli* to the royal court. Although King Pagan Min's government was paralysed with fear and indecision, the ordinary people refused to accept their new foreign master. For a decade, what was left of the Burmese army fought a guerrilla-style war against British forces in the south.[26]

In northern Burma, a new king, Mindon, succeeded the now insane Pagan Min, his half-brother. Fortunately, King Mindon (Plate 1) was a reasonable and peace-loving man, determined not to provoke the British further. In an effort to reverse Burma's misfortunes, he built a new royal capital, Mandalay, which was to become a centre of Buddhist studies and Burmese culture. King Mindon was genuinely interested in the welfare of his people and attempted to build up the economy of land-locked Upper Burma by encouraging trade with the British and introducing small-scale industry.[27]

In spite of reforms, King Mindon proved to be a traditionalist in the so-called 'shoe question'. Court etiquette demanded that all visitors remove their shoes before entering the royal palace. This practice had been imposed and, for the most part, accepted without objection since the earliest contacts with the West, but matters came to a head in the late 1870s when Burmese pride and provincialism clashed with imperial arrogance. To the British, representatives of the world's greatest empire, walking around in their stockinged feet was humiliating and insulting. After a bitter exchange of words and letters, the British Foreign Office instructed diplomats that, under no circumstances, should they comply with the rule. Royal officials also stubbornly held their ground.

1. King Mindon. (Yule, *Mission to the Court of Ava*, 1857)

Consequently, the king stopped meeting with the British Resident and business between the two governments was conducted by lower-ranking diplomats. Diplomatic relations between the two countries suffered yet another set-back.[28]

Aside from the public furore, many diplomats privately admired King Mindon's pluckiness. The king was, after all, just trying to save face for the Burmans. Unfortunately, King Mindon died in 1878 only a year after the heir apparent had been murdered. Once again there was a struggle for the crown, and Thibaw (r. 1878–85), Mindon's son although not by the chief queen, managed to win. A vacillating man, lacking in strength, judgement, and vision, King Thibaw was controlled by his wife, Queen Supayalat, a cunning, ruthless woman. It was Supayalat and not Thibaw who orchestrated a massacre of eighty members of the royal family. British merchants throughout Burma, Upper and Lower, viewed the situation with increasing alarm and urged Lord Dalhousie in India to resolve the situation once and for all by deposing the new king and annexing the north.[29]

Taking advantage of the situation, the Shans and Kachins, hill dwellers in the far north of the country, rose in rebellion against Mandalay. Rumours reached Calcutta that in return for much-needed military aid, King Thibaw was considering granting large concessions, such as a lease on ruby mines and the licence to operate steamers on the Irrawaddy River, to the French. After lengthy negotiations, the French decided not to get themselves caught in a confrontation between Britain and Burma. With the French out of the picture, the opportunity to intervene presented itself to the British when King Thibaw cancelled the Bombay Burmah Trading Company's long-held logging rights in the rich teak forests north of Toungoo. When the king refused to reconsider the matter, the Third Anglo-Burmese War began. It was a short engagement; the British occupied Mandalay on 14 November 1885. There was no resistance to speak of. King Thibaw and Queen Supayalat were quietly deposed and sent into exile in India.[30]

On 1 January 1886, the British annexed Upper Burma. In response, the usually passive natives rose up in fury. Earlier defeats had been humiliating, but defeat in the Third Anglo-Burmese War resulted in the loss of Burma's independence; the final blow to national pride and Burmese culture had been delivered.[31] The people did not meekly submit to subjugation. Burmese resistance forces kept 32,000 British troops and 8,500 military police occupied for five years before the British finally pacified Burma.[32]

Burma was a British colony for over sixty years, from 1886 until World War II. Trade dictated the choice of Rangoon as the colonial capital. Founded in 1755 by King Alaungpaya, the city was completely rebuilt by the British and reached its height as a great seaport, part of the vast Victorian Empire. Following independence in 1943, Rangoon, known today as Yangon, became Burma's national capital.

Burma's history from the glory of ancient Pagan to World War II is a complex, often violent, endlessly fascinating story of the creation of a nation. Military power, ethnic strife, oriental kings, and devotion to Buddhism form the main themes. Significant events surrounding the creation and destruction of the country's three Burmese empires occurred in three cities—Pagan, Mandalay, and Rangoon. We begin our journey through Burma's history along the Irrawaddy River on the dusty plains of Pagan.

1. Michael Aung-Thwin, *Pagan: The Origins of Modern Burma*, Honolulu: University of Hawaii Press, 1985, pp. 20–1. See also Arthur P. Phayre, *History of Burma: Including Burma Proper, Pegu, Taungu, Tenasserim, and Arakan*, 1883; reprint edn., New York: Augustus M. Kelley Publishers, 1969, pp. 1–23 and Nicholas Tarling, *A Concise History of Southeast Asia*, Singapore: Donald Moore Press Ltd., 1966, pp. 3–8.

2. D. G. E. Hall, *A History of South-East Asia*, 4th edn., London: Macmillan Education Ltd., 1981, p. 36. See also Charles Lee Keeton, *King Thibaw and the Ecological Rape of Burma*, Delhi: Manohar Book Service, 1974, pp. 1–3 and Phayre, *History of Burma*, pp. 28–9

3. Keeton, *King Thibaw*, pp. 1–3.

4. HRH Prince Damrong Rajanubhab, *Journey through Burma in 1936*, trans. Kennon Breazeale, Bangkok: River Books, 1991, p. 162; Hall, *History of South-East Asia*, p. 152.

5. Tarling, *Concise History of Southeast Asia*, p. 14.

6. G. H. Luce, *Phases of Pre-Pagan Burma: Languages and History*, 2 vols., London: Oxford University Press, 1985, Vol. I, pp. 47–8; Hall, *History of South-East Asia*, pp. 153–7; Phayre, *History of Burma*, pp. 18–19.

7. Pe Maung Tin and G. H. Luce (trans.), *The Glass Palace Chronicle of the Kings of Burma*, London: Oxford University Press, 1923, pp. 29–64; Phayre, *History of Burma*, pp. 22–3.

8. Hall, *History of South-East Asia*, p. 158; Phayre, *History of Burma*, pp. 33–4.

9. Hall, *History of South-East Asia*, p. 160.

10. Phayre, *History of Burma*, pp. 37–9; Aung-Thwin, *Pagan*, pp. 22–4; Hall, *History of South-East Asia*, pp. 162–4.

11. Phayre, *History of Burma*, pp. 39–40; Aung-Thwin, *Pagan*, p. 24; Hall, *History of South-East Asia*, pp. 164–5.

12. Phayre, *History of Burma*, pp. 49–54; Hall, *History of South-East Asia*, pp. 167–70.

13. Phayre, *History of Burma*, pp. 90–5.

14. Ibid., pp. 100–19.

15. Ibid., pp. 125–9; Hall, pp. 414–16. According to Phayre (p. 129), De Brito, who had destroyed pagodas in his thirst for treasures, 'was impaled on a high stake before his own house, and so lived for three days in dreadful agony'.

16. Phayre, *History of Burma*, pp. 134–6.

17. Ibid., pp. 149–57; Damrong, *Journey through Burma*, p. 29; Hall, *History of South-East Asia*, p. 427.

18. Phayre, *History of Burma*, pp. 168–9; Hall, *History of South-East Asia*, pp. 433–4.

19. Phayre, *History of Burma*, pp. 190–1; Hall, *History of South-East Asia*, pp. 434–7.

20. Phayre, *History of Burma*, pp. 209–11. See also Sterling Seagrave, *Dragon Lady: The Life and Legend of the Last Empress of China*, New York: Vintage Books, 1993, p. 68. Interestingly, Seagrave describes a similar practice in seventeenth-century China.

21. Hall, *History of South-East Asia*, pp. 625–35.

22. Phayre, *History of Burma*, pp. 210–15.

23. Hall, *History of South-East Asia*, pp. 646–8.

24. Hall, *History of South-East Asia*, pp. 648–52. See also J. G. Scott, *Burma from the Earliest Times to the Present Day*, London: T. Fisher Unwin, Ltd., 1924, p. 269.

25. Hall, *History of South-East Asia*, pp. 649–50.

26. Ibid., pp. 650–1; Tarling, *Concise History of Southeast Asia*, p. 146.

27. Hall, *History of South-East Asia*, pp. 666–71; Damrong, *Journey through Burma*, pp. 78–9; V. C. Scott O'Connor, *Mandalay and Other Cities of the Past in Burma*, London: Hutchinson and Co., 1907; reprint edn., Bangkok: White Lotus Co., 1987, pp. 4–9.

28. Scott, *Burma*, p. 296; see Seagrave, *Dragon Lady*, p. 67, regarding a similar problem in China.

29. O'Connor, *Mandalay and Other Cities*, pp. 20–4.

30. Keeton, *King Thibaw*, pp. 286–90; Hall, *History of South-East Asia*, pp. 676–81.

31. Aung San Suu Kyi, *Freedom from Fear and Other Writings*, ed. Michael Aris, Harmondsworth: Penguin Books, 1991, pp. 84–5.

32. Hall, *History of South-East Asia*, p. 771.

2
Pagan: City of Kings

PAGAN, Burma's first empire, endured for over four and a half centuries from its origins in the ninth century through unification to the final battle in the thirteenth century when Pagan succumbed before the mighty Kublai Khan. The empire's history was rich and complex, characterized by ruthless, semi-divine kings, mostly docile and obedient subjects, and nearly unsurpassed military power. What clearly distinguished Pagan, however, was the influence of Theravada Buddhism, which moulded society, motivated the actions of its kings, and, ironically, aided its decline.

The legendary King Pyuminhti established Pagan in 849 on the site of a former Pyu settlement not long after the Pyus had been overrun by invaders from present-day western Yunnan, China. Pyus were among the early inhabitants of Pagan; both ruins and metal artefacts clearly reveal traces of their workmanship. For over a century, Pagan's existence went unnoticed. The city was not mentioned in contemporary inscriptions until the accession of King Anawrahta to the throne in the eleventh century, first by the Chinese in 1004 and later by the neighbouring Chams in 1060 and the Mons in 1093. Today ruins of the city's walls and the Soraba Gate (Plate 2) remain, the only vestiges of this once fabulous city. Outside the walls lay the astonishing ruins of some 4,000 religious monuments spread out over approximately 40 square kilometres (Colour Plate 1). Yet these represent only a small fraction of the 13,000 structures said to have existed in the thirteenth century on the eve of the invasion of Kublai Khan's forces. Historians believe that some may have been hurriedly torn down and their bricks and timber used to fortify the besieged city. Many more have been reduced to rubble by earthquakes or simply washed away by the changing course of the Irrawaddy River.[1]

Exactly when temples and stupas began to appear on the dry, dusty Kyaukse Plain is as uncertain as the empire's early history. Only with King Anawrahta's accession in 1044 did Burma's history begin to come into focus. Even then facts were liberally mixed

2. The Soraba Gate, Pagan. (O'Connor, *Mandalay and Other Cities*, 1907; Antiques of the Orient, Singapore)

with legend. By all accounts, King Anawrahta was the stuff of which gods are made. His courage on the battlefield was unmatched, his tactics inspired, his religious ardour truly divine. We do know for certain that it was he who first unified Burma, extending Pagan's hegemony north almost to the border with China and south to the Isthmus of Kra. Contacts reached across the Salween River to the Shan Plateau and west as far as Arakan.[2]

King Anawrahta's greatest conquest of all was Thaton, the kingdom of the Mons in the south of Burma. He was primarily motivated by the desire to reform and purify a corrupt Buddhist clergy which had become too powerful and thoroughly dissolute, drawn to tantric rituals and animism. The situation was truly scandalous. The monks 'blatantly behaved in ways forbidden to the Buddihist Order, savouring the delights of both the bottle and the flesh'.[3] King Anawrahta launched a military campaign against the Mons only after peaceful efforts to acquire the Tripitaka, the Theravada scripture, had repeatedly been rebuffed.

King Anawrahta was also prompted to military action by Pagan's need for labourers to work on the many irrigation projects so necessary to the sustainment of life on the semi-arid plain. The war yielded some 30,000 Mon prisoners, many of whom were put to work on the state's farms, which then comprised over 28,000 acres of padi. Others worked to extend the kingdom's agriculture into the fertile Minbu Plain south-west of Pagan.[4]

Even as the hatred of the Mons increased, the victorious king and his court ironically fell under the spell of the Mons' culture, which they perceived as richer and more advanced than their own. Consequently, artisans, priests, scribes, and scholars were taken as prisoners to complement King Anawrahta's court. For nearly a century, Mon culture dominated Pagan's art, architecture, and politics. Pali, the language of the Theravada Buddhists, became the court language. Most significantly, a young Mon monk, Shin Arahan, attained a position of unparalleled influence at court.[5]

King Anawrahta and his courtiers enthusiastically embraced Theravada Buddhism which, in contrast to the more complex Mahayana Buddhism, taught that salvation could be achieved through good deeds, ranging from simple acts of charity such as filling the bowl of a begging monk to the construction of a pagoda or temple. The king consequently embarked upon a flurry of building projects to earn religious merit for himself and his subjects.[6] His humble and earnest devotion is revealed in the inscription on a pagoda:

By this abundant merit I desire
Here nor hereafter no angelic pomp
Of Brahmas, Suras, Maras; nor the state
And splendours of a monarch; nay, not even
To be the pupil of the Conqueror.
But I would build a causeway sheer athwart
The river of Samsara, and all folk
Would speed across thereby until they reach
The Blessed City.[7]

The Early Period of Architecture, 850–1120

In his zeal to save himself and his subjects, King Anawrahta built a great many pagodas or stupas, all of which are considered part of the Early Period of religious architecture in Pagan, which lasted from 850 to 1120. These solid monuments, which sprang up across the plain like a veritable forest of strange trees, featured central vaults to hold precious relics of the Buddha. Not surprisingly, monuments from this period reveal Indian and Pyu influence in design as well as evidence of Hinduism, influences which gradually gave way, as the Early Period progressed, to new indigenous styles in art and architecture and the influence of Theravada Buddhism.

One of the oldest shrines at Pagan, most likely dating from the middle of the tenth century, is the Nat-hlaung-kyaung, the Shrine of the Sacred Devas. Dedicated to Vishnu, it was a place where Brahman priests probably presided over Hindu rituals. Its bulbous shape is similar to the style of Pyu stupas.[8]

In contrast is the Shwe-hsan-daw Pagoda, built by King Anawrahta early in his reign to hold a strand of the Buddha's hair seized during a successful raid on Thaton. The fact that the king had been able to lay hands on the precious relic was taken as a sure sign that the Buddha himself approved of Pagan's campaign against the Mons. The pagoda's architectural design reflects a break with the past. Unlike the earlier bulbous Pyu stupas, the Shwe-hsan-daw has five terraces and is shaped like a pyramid. Its location was also important; the Shwe-hsan-daw was one of four pagodas which marked Pagan's city limits, following a custom at the Pyu capital of Sri Ksetra.[9]

King Anawrahta also built a library, the Pitaka-taik, to house the thirty sets of the Tripitaka brought in triumph to Pagan from Thaton. This square monument featured a central chamber to hold the palm-leaf manuscripts safe from the sun's rays, the monsoon rains, and the threat of fire. Pilgrims could gaze on the sacred books from an outer ambulatory through small perforations in the thick walls. Scenes from the Jatakas arranged in two tiers along the library's outer walls silently instructed the faithful about various events in the life of the Buddha.[10]

King Anawrahta's final monument was the Shwe-zigon Pagoda (Colour Plate 2), which he began and his successor King Kyanzittha completed. The Shwe-zigon Pagoda, whose name means Land of Victory, was built to hold bones and a tooth of the Buddha. The monument's massive style is strongly reminiscent of Pyu architecture, but there are also indications of a budding indigenous culture apparent in the pagoda's four bronze Buddhas, which date from about 1102, and were most certainly the work of Burmans and not craftsmen from India. Undoubtedly, the most interesting feature of the Shwe-zigon (one which also points to an emerging Burmese culture) is a set of thirty-seven beautifully carved statues of *nat* (the spirits of people who had died untimely deaths and whose souls roamed the earth seeking justice from their abode on Mount Popa south-east of Pagan). These were originally erected on the lower terraces but later moved to a hall south-east of the pagoda. A statue of the Hindu god Indra, known to the Burmans as Thagyamin, Chief God of the *Nat*, also graces the temple, the only such image in Pagan. From tribal days, the Burmans had worshipped *nat*, who, being more often than not malevolent, demanded endless propitiation. In an effort to gain control of the old animistic beliefs, King Anawrahta limited the number of *nat* to thirty-seven spirits whose function it was to protect the royal family and the state. Of course, the *nat* never lost their hold on the national consciousness, and in the end, Buddhism in Burma accommodated *nat* worship. Today, while most of Pagan's pagodas and temples have either fallen into ruin or attracted visitors interested only in their architectural merits, the Shwe-zigon remains a national shrine as vibrant as it must have been almost 1,000 years ago.[11]

Life in Pagan

Following victory over the Mons, Pagan experienced a period of considerable upheaval caused in part by a great influx in population. At the height of its power, Pagan's population is estimated to have numbered nearly 2 million, including Burmans, Mons, Shans, Karens, Indians, Arakanese, and other smaller groups. The traditional structure of Pagan's society, at once highly centralized and rigidly hierarchical, acted to ensure order and stabilize society as the newcomers were assimilated. The royal family was at the pinnacle of society followed by court officials, both higher and lower. A second tier contained artisans and merchants who served the crown. Finally, at the base of the pyramid were the commoners, most of them agricultural workers, who made up the bulk of society and who had neither political nor economic power.[12]

Parallel to but separate from Pagan's secular society was the Buddhist clergy, the sangha, which played an important role in serving as a conduit to the Buddha, offering prayers and sacrifices on behalf of the lay people. At the top was the primate or Mahasangharaja, a position held for many years by Shin Arahan. Directly below the primate were the provincial abbots who supervised monasteries throughout the land. Finally, there were the village priests, ordinary monks, and novices at the bottom. With shaven heads and simple yellow robes, monks and novices silently moved about the people daily begging for food (Plate 3).[13]

For the most part, little is known about the lives of the ordinary people of Pagan. Life in an agrarian society was an endless cycle of seasons with the wet monsoon beginning in May lasting through most of October followed by a cooler, drier period from November through February, and a truly dry season in March and April. Religious festivals marked the progress of the seasons, helping to break the tedium of everyday life. One event, celebrated in Burma every year until the end of the monarchy in the nineteenth century, was Le Dun or the royal ploughing (Colour Plate 3). In a colourful, carefully orchestrated pageant dictated by centuries of tradition, and designed to bring a bountiful harvest, the king and his court took to the field to plough several strips behind a team of white oxen richly arrayed in jewelled harnesses.[14]

It is difficult if not impossible for late twentieth-century

3. Novices returning to their monastery with their alms bowls. (Singer, *Burmah*, 1993)

Westerners to comprehend how Pagan's society functioned, given our concept of individual rights as a condition of humanity. Citizens of Pagan existed not as individuals but as members of groups, a membership based not on rights and freedoms but on obligations. Heredity determined everything from where a person lived to his occupation and obligations to the crown and society. Mons, Shans, and other ethnic groups lived in separate villages named accordingly. Burmans were divided by occupation. Masons, sculptors, and locksmiths lived in their own villages. In addition, certain occupations were hereditary. The sons of cavalry officers, brick masons, scribes, potters, and *pura kywan* (a term incorrectly but commonly translated as temple slaves) had no choice but to follow in their father's footsteps. In this way, Pagan's government attempted to maintain control of labour and to guarantee an adequate number of skilled workers.[15]

Bondage, voluntary and involuntary, combined with heredity and ethnicity to hold the fabric of Pagan's society together. Peasants, or *kywan*, freely pledged their services to the government, to a wealthy patron, or to the sangha in return for wages in kind. Peasants bonded to a landowner, for example, received a part of the harvest in return for their labour. When a wealthy man pledged money or land to the clergy (a common occurrence), he generally included the services of one or more peasants in perpetuity. Those pledged were known as *pura kywan* and their bondage became hereditary. Should a man become indentured because of debt, his burden also passed on to his heirs.[16]

In his book, *Mandalay and Other Cities of the Past in Burma*, published in 1907, V. C. Scott O'Connor tells of meeting Prince Mintha, the last descendant of the Mons captured by King Anawrahta, who was head of the *pura kywan*. According to the prince, it was quite common for nobles to donate entire villages to the service of the sangha, often in perpetuity.[17] In 1223, for example, a wealthy woman of Pagan stipulated that 'having no father or heir to inherit my estate I dedicate on this the 11th waning Pyatho, 585 Sakkaraj, Monday, the whole of the villagers of Nga Myindaung, for the upkeep of the recumbent image on the eastern side of my cave and of the image erected by monk Pwa-gyi'.[18]

Another inscription dating from 1271 reveals the generosity of a patron who shared the merit which he earned with the very slaves he was donating:

> On Thursday, the 5th waning of Thadingyut, 633 Sakkaraj, Nga Pyinnya Thin dedicated his personal slaves Nga Ya Kaing and Nga Thaung Ga in order that he and they might escape the horrors of Avici hell. The slaves willingly submitted to this dedication while they were resting their heads on Nga Pyinnya's loins at a place called Ngataba.[19]

Frequently, people bought slaves for the sole purpose of handing them over to the clergy. For example, a man by the name of Yattapisi and his wife dedicated eighteen people 'bought with the produce of our own labour and not inherited from our parents'. Perhaps the most impressive inscription of all tells of a man so devout that he dedicated himself, his wife, and two sons as pagoda slaves.[20]

The ranks of the *pura kywan* contained not only prisoners of war and common labourers but also skilled artisans. At one time, there apparently was a certain degree of status associated with the position. Later, however, because of their constant contact with lepers and other diseased persons, the pagoda slaves became outcasts, pariahs of the very society which they were condemned to serve. They had, in fact, become non-persons completely lacking in social status.[21]

King Anawrahta died in 1077, after ruling Pagan for over thirty years. He left behind a multicultural society divided by a variety of religions, old and new, a babble of tongues, and a variety of traditions and daily lifestyles. Much to his credit, his successor, King Kyanzittha, was able to balance the various elements and attain a certain degree of harmony. According to one Burmese scholar, King Kyanzittha's reign was characterized by 'Burman military rule, Pyu traditions, Mon culture, and Theravadin [*sic*] spirit'.[22]

King Kyanzittha's coronation in 1086 was both spectacular and symbolic. Appealing to the entire kingdom, he employed rituals of the Hindus, Buddhists, and *nat* worshippers as well as a mix of Burmese, Mon, and Pyu hymns. Keeping with tradition, he accepted the five regalia, which included a white umbrella, yaktail fan, crown, sword, and sandals. The new king rode through Pagan on a white elephant reserved exclusively for his use. Most significantly, King Kyanzittha foretold his coming Buddhahood, repeating the words attributed to the Buddha himself: 'I am the foremost in all the world.' Although he was the first Burmese king to make such a prophecy, there is abundant evidence that rulers elsewhere in the region made similar claims to divinity.[23]

Eleventh-century Pagan was a splendid place which symbolized Tavatimsa, the centre of the universe. The walled city had twelve gates, each bearing a sign of the zodiac. The four main gates were located at the points of the compass. At intervals along the wall, there were thirty-two parapets, each with a seven-tiered roof, symbolizing the royal *nat*. Four large pavilions rose above the main gates, again each with seven tiers. Precisely in the centre of the city was the largest pavilion of all, rising majestically above the Lion Throne, the very centre of the universe. Destined as he was to become a Bodhisattva or a *nat*, depending on one's point of view, the semi-divine occupant of the Lion Throne was the earthly manifestation of Sakka, king of Tavatimsa. The king of Pagan was therefore responsible for the very salvation of his people.[24] Pagan's physical plan (Figure 2), loaded as it was with symbolism, was faithfully copied in Burma's royal capitals right down to the construction of Mandalay in the nineteenth century.

During King Kyanzittha's reign, Pagan embarked upon its greatest period of temple building, another symptom of an expansive royal ego. The king was responsible for many of Pagan's most notable temples, including the Ananda Temple, his greatest achievement of all, which dates from around 1105 (Colour Plate 4). According to legend, eight Indian monks appeared one day at the palace begging for alms. Captivated by their description of a Buddhist shrine in a cave in the pristine wilderness of the Himalayas, King Kyanzittha determined to build a replica of it on the Pagan Plain. Everything about the Ananda, the first of Pagan's great temples, is grand, from its exterior and arched gates to its wonderful symmetry (Figure 3). Here is evidence that Pagan had indeed 'developed her own architectural dialect from a combination of indigenous prototypes and the latest of North Indian artistic fashions'.[25] While evidence of Indian workmanship is abundant throughout, the basic plan is that of a cross taken from earlier Pyu designs.

Built to instruct as well as to inspire, the Ananda's exterior bears

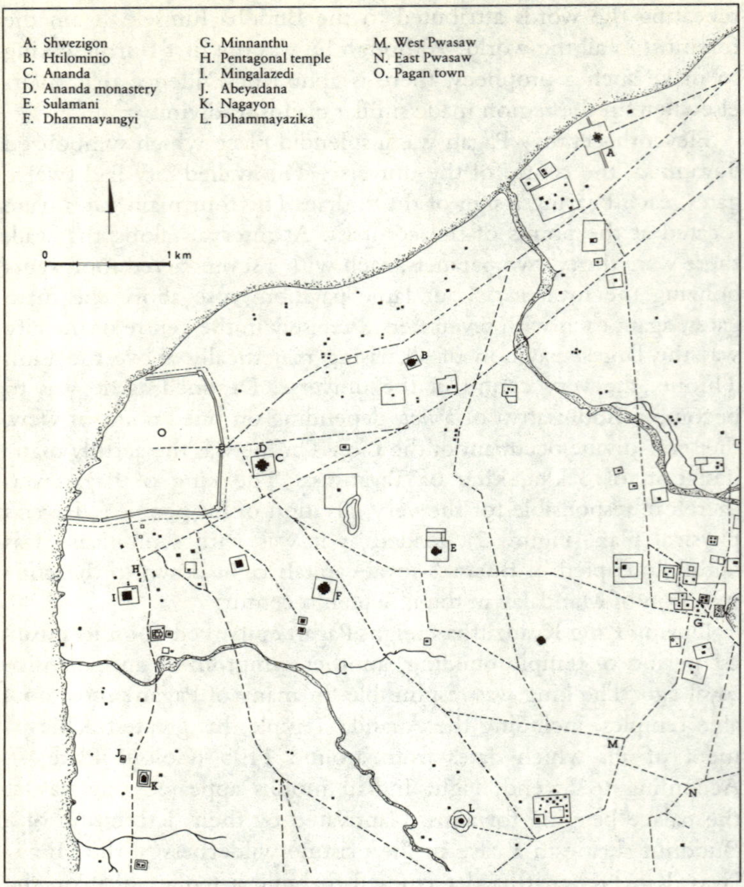

Fig. 2 Plan of Pagan. (Courtesy of Jacques Dumarçay)

a series of 554 panels, which decorate the temple's plinth at or slightly below eye-level and depict scenes from the defeat of the Army of Mara by the Buddha at the supreme moment of Enlightenment. The Buddha is represented in the form of a stupa because it is disrespectful to place an image of the Buddha below the level of one's body. The terraces, both lower and upper, were also utilized for instructing the faithful. In all, there are slightly over

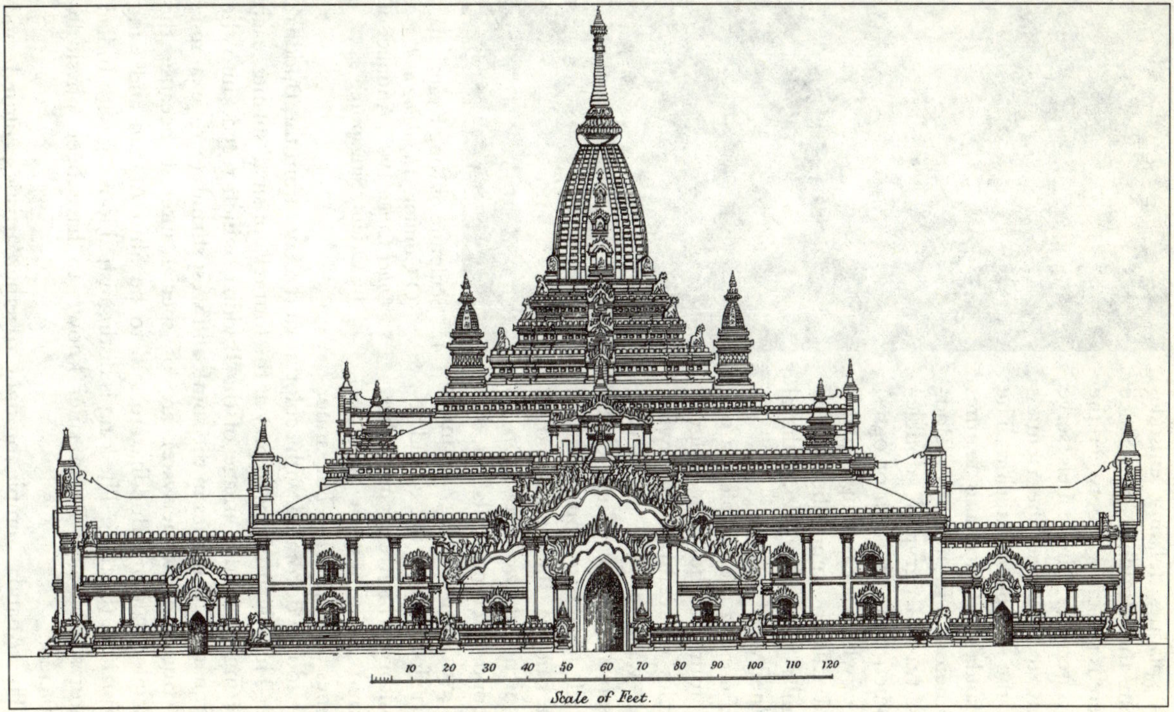

Fig. 3 Elevation of the Ananda Temple, Pagan. (Yule, *Mission to the Court of Ava*, 1857)

900 plaques relating episodes from the Buddha's life. Since there is no stairway between the two terraces, the panels serve only to enhance the richness of the temple's exterior. The Ananda's exterior also features numerous figures of Bodhisattvas, 'leogryphs', double-bodied lions, and *bilu* or ogres, all of which act as temple guards and call to mind the use of similar figures in the medieval cathedrals of Europe.[26]

4. Sculpture of King Kyanzittha. (O'Connor, *Mandalay and Other Cities*, 1907; Antiques of the Orient, Singapore)

Entering the temple via one of four porches, the pilgrim circles the central mass twice. He walks slowly meditating upon a series of reliefs, eighty in all, again based on the life of the Buddha, the Jataka, and set in niches at two levels. On his second circumambulation, the pilgrim walks along an inner corridor next to the closed repository which contains relics, sacred scriptures, or precious stones and jewels. The inner corridor is decorated with sculpture and scenes from the Buddha's life. The pilgrim stops to pray at the four *cella* or small shrines, each bearing a wooden image of the Buddha, which recede into the reliquary. The statues in the south and west shrines (early examples of Burmese lacquerware) are of special interest because the disciples flanking Buddha are said to be Shin Arahan and King Kyanzittha (Plate 4). The king is distinguished looking with Indian features, the only Pagan ruler known to have been physically depicted.[27]

The Ananda Temple brought together artistic, cultural, and

religious currents to reflect a society which was both multicultural and harmonious, a society very much at ease with itself. Pyu techniques are combined with elements of the more advanced art and theology of the Mons. The temple conveys a strong sense of indigenous styles melded with outside influences from India to create a monument that was uniquely Burmese.

The Middle Period of Architecture, 1120–1170

King Kyanzittha's reign marked the springtime of Pagan's youth and was followed by the full bloom of Alaungsithu's summer. Exuberance, self-confidence, a spirit of optimism, and, most of all, an overwhelming sense of nation-hood pervaded the Middle Period of the Pagan Dynasty. During King Alaungsithu's half-century reign, from 1112 to 1167, Pagan's architecture entered a new phase in which interiors became brighter and exteriors attained new heights in grace and beauty. Increasingly, during this period, Burmese replaced Pali in inscriptions.[28]

In comparison to the military might of King Anawrahta and the cultural achievements of King Kyanzittha, King Alaungsithu's accomplishments might seem pedestrian and uninteresting at first glance. Yet it was Alaungsithu, the grandson of Anawrahta, who made certain that the gains of his predecessors would not be lost to the whims of a future but less capable ruler. Known as the 'canal digger', he was an effective administrator who standardized the bureaucratic, political, and economic systems which made Pagan long-lasting.[29]

One of King Alaungsithu's earliest projects, the Shwe-gu-gyi Temple, was constructed within a period of seven months in the year 1131. The temple is unusual in two ways for it is located within the city walls, and it sits upon a tall platform. From its tapered corner obelisks to its tall *sikhara*, the monument's overall design gives it a sense of upward thrust which is at once graceful but restrained. The floor plan also differs from earlier temples in that there are no ambulatories and the inner shrine is located directly below the *sikhara*. Many critics find the shrine's plain, whitewashed interior disappointing in contrast to the 'supreme competence and originality of the exterior'.[30]

The inscription of the Shwe-gu-gyi, supposedly written by the king, a devout Buddhist, raised inscriptions to new poetic standards.

But I would build a causeway sheer athwart
The river of Samsara, and all the folk
Would speed across thereby until they reach
The Blessed City.
I myself would cross
And drag the drowning over
Ay, myself tamed, I would tame the wilful;
Comforted, comfort the timid;
Wakened, wake the asleep;
Cool, cool the burning;
Freed, set free the bound,
Tranquil and led by the good doctrines
I would hatred calm.[31]

King Alaungsithu's architects best captured the nationalistic spirit of the age in the That-byin-nyu Temple, 'the Omiscient', built near the end of his reign. In the words of one critic, 'it is truly a monument to this phase in Burma's history when a spirit of innovation and experiment led to great climaxes of creative energy that were an expression of the self-confident Burmese spirit of nationhood'.[32] The That-byin-nyu is composed of two massive cubes and, like the Shwe-gu-gyi, is located within the city walls about 500 metres south-west of the Ananda. What is especially striking about this monument (besides the fact that it is the tallest in Pagan at 61 metres) is that the main shrine is on the upper level. The visitor enters at ground level and climbs to the upper level via an interior staircase. What greets him at the top is the inner sanctum and beyond a wonderful panorama of Pagan. The experience conveys both the majesty of earthly kings who built the hundreds of shrines on the plain below and the greater power and compassion of the Buddha. 'This is no lifeless pile of brick and plaster,' writes one admiring scholar. 'It is the Purha, the Buddha, exhibiting the great miracle. See the great miracle. See the deluge descending from his shoulders! See the holocaust ascending from his feet.'[33]

One last monument from this period which fascinates historians and tourists alike is the Dhamma-yan-gyi Temple (Plate 5). Not long after its completion, the temple's arches and windows were

5. The Dhamma-yan-gyi Temple, Pagan. (Yule, *Mission to the Court of Ava*, 1857)

sealed off, leaving only the east shrine open. The culprit will never be known. Only the motive is clear: to deny the eternal merit thus earned by the royal builder for himself and his subjects. One account, favoured by many scholars, contains all the elements of a Shakespearian tragedy. According to this version of events, King Alaungsithu built the temple at the end of his long reign, and shortly after became deathly ill. Unable to wait for natural death, Narathu, the king's son and heir, took the old man to the Shwe-gu-gyi Temple where he smothered him to death. That, however, was not enough to assuage the hatred of the son for the father. The new king sealed off the Dhamma-yan-gyi Temple ensuring that no prayer would ever rise to heaven from its sacred grounds in his father's name.[34]

The Late Period of Architecture, 1170–1300

During the century after King Alaungsithu's death, Pagan continued to grow and prosper, but the seeds of its destruction had clearly taken root. King Narapatisithu (r. 1173–1210) and his successor, King Nantaungmya (r. 1210–34), expanded the nation to its largest territorial limits.[35] Unfortunately, donations of both land and labour to the sangha increased significantly and offset the territorial gains. The wealth of the kingdom and, with it, the balance of power were being transferred from the king to the clergy.[36]

During the Late Period, religious architecture continued to develop its 'distinctive Burmese style'. Temple interiors became brighter in colour while exteriors took on more upward thrust. In addition, a new pentagonal floor plan was implemented in some monuments.[37] In a frenzy of construction, over 1,000 monuments were built within the relatively short span of 130 years. One temple from this time period, the Sula-mani, is acknowledged as Pagan's 'ultimate temple', because it became the prototype for later kings who copied it again and again. The Sula-mani 'ideally expressed all that a royal act of merit should: a demonstration of imperial prestige and stately grandeur and a glorious confirmation of faith in the world religion and of royal support for it'.[38] Built by King Narapatisithu in 1181, the temple combines a massive square base with a pyramidal-shaped upper shrine which gracefully reaches upward finishing in a tall, slender *sikhara*. Its interior is bright with colour and openness. Also noteworthy is the total absense of Pali which had been replaced by Burmese.[39]

King Narapatisithu, whose reign spanned nearly forty years, was also responsible for the Dhamma-yazika Temple situated some distance from the city walls. Completed in 1196, when the empire was still at its height, this was the first temple to feature a pentagonal floor plan (Figure 4), an architectural innovation that signified a substantive change in Buddhist theology. The addition of a fifth side symbolized the acceptance of a fifth future Buddha, Mettaya. Inscriptions indicate that Mettaya's cult was extremely popular with the king and his court as well as the common people, all of whom hoped to be alive to welcome the 'benefits of salvation at the hands of an actual Buddha who would offer mankind a dispensation'.[40] The ruins of a large number of ancillary buildings indicate that the temple attracted large numbers of pilgrims.

The appearance of pentagonal monuments outside the city in no way eclipsed the traditional four-sided temple. Indeed, what has been described as the 'most refined' of Pagan's temples, the Gadaw-palin (Plate 6), was built on the traditional plan at the end

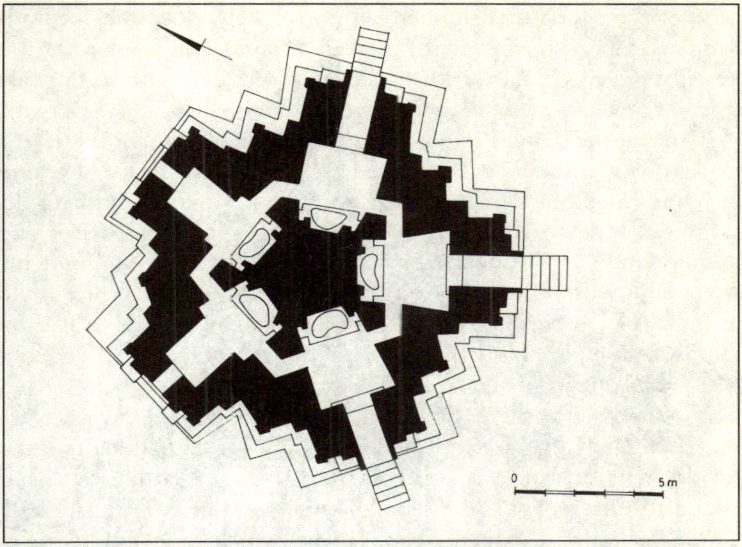

Fig. 4 Plan of a pentagonal monument, Pagan. (Courtesy of Jacques Dumarçay)

of King Narapatisithu's reign. While Pagan's many temples seem to compete for superlatives among art historians, the Gadaw-palin remains the favourite of many visitors. Here, on the banks of the Irrawaddy River, Pagan's kings paid homage to their ancestors. Many of the temple's features are quite traditional, from its upper level main shrine to the four shrines on the ground floor facing the cardinal directions. What is so appealing about the Gadaw-palin is its elegance and grace. Finding it too romantic, later rulers avoided copying the Gadaw-palin, preferring instead to build temples 'projecting an impression of stateliness and composure'.[41] In other words, the ordinary or the stereotyped prevailed over the unique and the unexpected.

Architectural innovations and creativity belied Pagan's decline. By the mid-thirteenth century, Buddhist temples were 'gorged with wealth', a symptom of the ongoing transfer of wealth from crown to the clergy. Monasteries sprang up throughout Pagan, and no ruler was strong enough to challenge the powerful sangha.

6. The Gadaw-palin Temple, Pagan. (Yule, *Mission to the Court of Ava*, 1857)

Moreover, the undisciplined clergy split into a 'hotchpotch' of sects and was once again badly in need of reform. Asceticism was abandoned as even ordinary monks became 'involved in peasant uprisings, land disputes, land buying and selling, and throwing lavish feasts (consisting of whole cattle, pigs, and liquor) to celebrate legal victories'.[42]

Ignoring shrinking revenues, King Narathihapate (r. 1254–87), a tyrant devoid of religous devotion, continued to build temples.[43] His best-known project was the Mingala-zeidi Temple, completed ten years before the fall of Pagan. This large square stupa is famous for its full set of glazed plaques depicting the Jataka and manages to convey a strong sense of self-confidence. But it is only braggadocio masquerading as confidence. The stupa's inscription clearly reveals the depraved character of the man who led Pagan in its closing decade. King Narathihapate boasted that he was the 'supreme commander of a vast army of 36 million soldiers, the swallower of 300 dishes of curry a day'.[44] Burmese chronicles describe a thoroughly dissolute monarch who possessed 3,000 concubines and, in no uncertain words, place the blame on him for the country's downfall.

In 1287, the forces of Kublai Khan overran Pagan and sacked the royal capital after King Narathihapate had refused to pay tribute to the Mongol conqueror. Upon hearing that the Tartars were about to attack, the king fled south to Bassein, confirming that, in addition to being brutal and decadent, he was also cowardly.

The epoch of Pagan was finished, a great empire destroyed, collapsed upon itself. Its influence, however, has continued through Burma's history even to the present day. Pagan connects the nation through time linking seemingly diverse episodes into a single coherent story. Later dynasties traced their ancestors to Pagan in an effort to establish their legitimacy. Exactly what the kings of Ava, Amarapura, and Mandalay knew about Pagan is unclear. Beyond religious inscriptions, there were no written records to speak of. Nevertheless, it was important to claim that coronation rites, royal audiences, and other kingly traditions were identical to those of Pagan. Pagodas and temples sprang up throughout Burma, all modelled on prototypes at Pagan. Most important of all, Theravada Buddhism, the religion of Pagan, became the national religion of Burma. Over the centuries, Pagan took on a mythical aura. Its ruins came to symbolize the Golden Age of Burma. Pagan's place in the nation's consciousness was perhaps strongest in the mid-nineteenth century as the Konbaung Dynasty faltered and Burma's nationhood neared its tragic end.

1 Michael Aung-Thwin, *Pagan: The Origins of Modern Burma*, Honolulu: University of Hawaii Press, 1985, p. 21. This extremely useful book is one of the few works available dealing with Pagan's society and history. Most works concentrate on architecture. See also V. C. Scott O'Connor, *Mandalay and Other Cities of the Past in Burma*, London: Hutchinson and Co., 1907; reprint edn., Bangkok: White Lotus Co., 1987, p. 241.

2. D. G. E. Hall, *A History of South-East Asia*, 4th edn., London: Macmillan Education Ltd., 1981, pp. 157–8. Hall is encyclopaedic and particularly useful for comparing neighbouring countries.

3. HRH Prince Damrong Rajanubhab, *Journey through Burma in 1936*, trans. Kennon Breazeale, Bangkok: River Books, 1991, p. 164.

4. Aung-Thwin, *Pagan*, p. 191.

5. Hall, *History of South-East Asia*, pp. 158–9.

6. Aung-Thwin, *Pagan*, pp. 22–3, 191. On the subject of Buddhism, see Edward Conze, *A Short History of Buddhism*, London: George Allen and Unwin, 1980, pp. 82–3, 104–5. Conze points out that Mahayana Buddhism did not die out in Burma but continued to influence religious art and thought throughout Pagan's history.

7. Pe Maung Tin and G. H. Luce (trans.), *The Glass Palace Chronicle of the Kings of Burma*, London: Oxford University Press, 1923, p. xi.

8. Paul Strachan, *Pagan: Art and Architecture of Old Burma*, Gartmore: Kiscadale Publications, 1989, p. 39. For an enlightening discussion of Pagan's architectural features, see G. H. Luce, 'The Greater Temples of Pagan', in *Fiftieth Anniversary Publications*, No. 2, Rangoon: Burma Research Society, 1960, pp. 169–73.

9. Strachan, *Pagan*, pp. 42–4.

10. Ibid., p. 45. See also Michael Aung-Thwin, *Historical Sites in Burma*, Rangoon: Ministry of Union Culture, Government of the Union of Burma, 1972, p. 44.

11. Strachan, *Pagan*, pp. 57–8; Damrong, *Journey through Burma*, pp. 182–4; Aung-Thwin, *Historical Sites*, p. 44; F. K. Lehman (Chit Hlaing), 'Freedom and Bondage in Traditional Burma and Thailand', *Journal of Southeast Asian Studies*, XV (Sept. 1984): 243.

12. Aung-Thwin, *Pagan*, pp. 71–6. See also Michael Aung-Thwin, 'Kingship, the Sangha, and Society in Pagan', in Kenneth R. Hall and John K. Whitmore (eds.), *Explorations in Early Southeast Asian History: The Origins of Southeast Asian Statecraft*, Ann Arbor: Center for South and Southeast Asian Studies, University of Michigan, 1976, pp. 208–10 and Michael Aung-Thwin, 'Hierarchy and Order in Pre-colonial Burma', *Journal of Southeast Asian Studies*, XV (Sept. 1984): 224–32.

13. Aung-Thwin, 'Hierarchy and Order', p. 225.

14. O'Connor, *Mandalay and Other Cities*, p. 46 and plate opposite; Aung-Thwin, *Pagan*, p. 49.

15. Aung-Thwin, 'Kingship', pp. 208–9.

16. Aung-Thwin, 'Hierarchy and Order', pp. 226–8; Aung-Thwin, 'Kingship', pp. 208–20; Lehman, 'Freedom and Bondage', pp. 233–4. Lehman's article is useful for he sheds more light on the concept of bondage by comparing language and practices in Burma and Thailand.

17. O'Connor, *Mandalay and Other Cities*, pp. 285–7.

18. Ibid., p. 288.

19. Ibid.

20. Ibid., p. 290.

21. Ibid., pp. 290–2; Lehman, 'Freedom and Bondage', p. 236.

22. Aung-Thwin, *Pagan*, pp. 23–4.

23. Hall, *History of South-East Asia*, p. 162; Aung-Thwin, *Pagan*, pp. 49–50. See also George Coedès, *Angkor: An Introduction*, trans. Emily Floyd Gardiner, 1961; reprint edn., Singapore: Oxford University Press, 1986, pp. 21–33. Coedès discusses the idea of the cult god at length.

24. Aung-Thwin, *Pagan*, pp. 47–51.

25. Strachan, *Pagan*, p. 65; Pe Maung Tin and Luce, *Glass Palace Chronicle*, p. 10.

26. Strachan, *Pagan*, pp. 65–71; Aung-Thwin, *Historical Sites*, pp. 58–60.

27. Strachan, *Pagan*, pp. 69–70.

28. Aung-Thwin, *Historical Sites*, p. 65; Hall, *History of South-East Asia*, p. 164; Strachan, *Pagan*, p. 79.

29. Aung-Thwin, *Pagan*, p. 24; Aung-Thwin, 'Kingship', pp. 232–3; Pe Maung Tin and Luce, *Glass Palace Chronicle*, pp. 111–28.

30. G. H. Luce quoted in Strachan, *Pagan*, p. 82.

31. Translated by G. H Luce, quoted in Hall, *History of South-East Asia*, p. 165.

32. Strachan, *Pagan*, p. 84.

33. G. H. Luce quoted in Strachan, *Pagan*, p. 87. See also Luce, 'Greater Temples', p. 174.

34. See Strachan, *Pagan*, pp. 87–8 and Hall, *History of South-East Asia*, p. 165 for another version of events.

35. Aung-Thwin, 'Kingship', p. 254. Aung-Thwin notes that the kingdom's boundaries stretched from East Bengal to the east bank of the Salween River, north to the Nanchao border, and south to Mergui in Tenasserim.

36. Aung-Thwin, *Pagan*, p. 187.

37. Hall, *History of South-East Asia*, p. 167.

38. Strachan, *Pagan*, p. 96.

39. Aung-Thwin, *Historical Sites*, pp. 66–7.

40. Strachan, *Pagan*, pp. 121–4; Aung-Thwin, *Historical Sites*, p. 146; Pierre Pichard, *The Pentagonal Monuments of Pagan*, Bangkok: White Lotus Co., 1991, pp. 14–19. Pichard discusses the theory and significance of the pentagonal plan pointing out the difficulties involved in incorporating a fifth side when traditionally temples were laid out so that the shrines of the four Buddhas coincided with the four cardinal directions.

41. Strachan, *Pagan*, pp. 97–8.

42. Aung-Thwin, 'Kingship', pp. 229–30; Aung-Thwin, *Pagan*, pp. 26–8.

43. Aung-Thwin, *Pagan*, p. 194; Hall, *History of South-East Asia*, p. 169.

44. Strachan, *Pagan*, p. 102.

3
Mandalay: City of the Buddha

FOR over four centuries, Pagan's kings ruled Burma confident of their divine abilities and certain that the royal city represented the very centre of the universe. Subsequently, however, other Burmese dynasties lacked the self-confidence and longevity of the Pagan Dynasty. As if doubtful of their legitimacy, later kings carefully traced their ancestry to King Anawrahta. They modelled court ritual and national traditions on what they believed to have been practised at Pagan. Most importantly, their royal cities were laid out on the banks of the Irrawaddy River according to the plan of Pagan, and they built palaces, temples, and pagodas which imitated structures in the ancient city—all of this in a never-ending effort to convince the common people of the royal links which existed between an extraordinary past and an ordinary present.

In turn, the royal cities of Ava, Amarapura, and finally Mandalay mirrored Pagan. The construction of Mandalay, the last royal capital of Burma, beautifully illustrates this fixation on the past. Mandalay was the project of King Mindon, who succeeded to the throne of the Konbaung Dynasty as a result of a palace coup at Amarapura in 1853. Much to everyone's relief, the new king displayed none of the symptoms of insanity which seemed to plague the Konbaung Dynasty. A devout Buddhist and dedicated pacifist, he wanted nothing more than to restore peace, harmony, and pride in his troubled land.[1]

No matter how dedicated or talented, the king could do little to alter the dismal fact that Upper Burma was a mere shadow of the great nation forged by King Alaungpaya. King Mindon's domain was land-locked and given to frequent droughts. In the south, the enemy British controlled the rich delta lands as well as the port cities of Rangoon, Bassein, and Martaban. The lost territory resulted in a diminished sense of Burma's national pride, nowhere more in evidence than at the king's court at Amarapura, tainted as it was by military defeat and royal insanity. In an attempt to reverse his country's fortunes and enhance the dynasty's prestige, King Mindon decided to abandon Amarapura in favour of a new royal capital. The massive rebuilding of Rangoon by the British most likely also motivated him.[2]

The king's decision was not at all unusual in Burma's history. The capital city had, in fact, been relocated several times earlier in the Konbaung Dynasty, almost always in the wake of some crisis. The first Konbaung court was short-lived at King Alaungpaya's hometown of Shwebo 100 kilometres north-west of Mandalay.[3] In 1765, King Alaungpaya's son and successor, Hsinbyushin, moved the capital to Ava in order to gain control of the Kyaukse Plain, the heart of the Burmese homeland. The city had been built by the Shans in the mid-fourteenth century, faithfully following the plan of Pagan.[4]

King Hsinbyushin's successors remained in Ava for only a generation. In the spring of 1783 following a failed coup, King Bodawpaya built Amarapura about 6 miles north of Ava on the Irrawaddy River (Colour Plate 5). In both its layout and architectural features, Amarapura closely resembled the southern city of Pegu. Originally built by the Mons, Pegu was rebuilt in the mid-sixteenth century by King Bayinnaung of the Toungoo Dynasty. King Bayinnaung, like his successors, looked longingly back to Pagan as he remodelled Pegu. The new Pegu was almost a perfect square with the royal palace situated in the middle and a moat surrounding it. Pegu was larger than Pagan and was more heavily fortified. Its arched gateways, symbols of royalty, were repeated at Amarapura and later at Mandalay. Pegu was also known for its richly gilded buildings, another feature copied at Mandalay. The fame of the Toungoo capital was, in fact, so widespread that contemporary European books featured drawings of its plan. Pegu became the strongest link in the historical chain as the living model of Pagan. In addition to the city's attractive features, the Konbaungs found much to admire in King Bayinnaung, who had successfully united Burma and had even extended his rule to Chiang Mai and Ayut'ia in neighbouring Thailand.[5]

In 1819, shortly before the outbreak of the First Anglo-Burmese War, King Bodawpaya's grandson, Bagyidaw (r. 1819–37), acting capriciously, moved his court back to Ava. Just as the golden tee, the royal umbrella, was placed over the Lion Throne, a violent thunderstorm broke out and the umbrella was dashed to the ground. Immediately realizing the full impact of such an ominous sign, the king ordered the execution of the architect.[6]

Nothing, however, assuaged the angry spirits; King Bagyidaw's reign turned out to be disastrous. Burma suffered substantial wartime losses of territory to the British. King Bagyidaw personally lost a small fortune simultaneously fighting the war and building temples and pagodas in an effort to enhance his reputation with his people.[7]

In 1837, King Tharrawaddy (r. 1837–46) acted on the advice of his astrologers and moved the court back to Amarapura. According to the astrologers, King Bagyidaw was responsible for the defeat because he had moved the royal court south instead of north. Ava, the country's second capital, was located upriver from Pagan and Amarapura was still further upriver. King Bagyidaw, they charged, had backtracked against the tide of history and the country paid a heavy price for his error. In their haste to place the blame on the former king, the astrologers neatly ignored the sorry state of the military, a much more obvious reason for Burma's defeat.[8]

The well-known British diplomat, Henry Yule, attributed the frequent royal peregrinations to Burma's navigable rivers and 'the unsubstantial nature' of her native buildings.[9] Certainly, there was a measure of truth to Yule's argument. The teak and bamboo buildings of the royal court were, in fact, easily dismantled, loaded on to elephants and taken by boat to the new site. Nevertheless, more subtle factors seemed to have been at play. A native observer argued that the move was necessary because the city's guardian spirits had become impotent and were no longer able to protect the royal city. The spirits were the ghosts of those buried alive under the city's gates and four corners. They were charged with the task of guarding the king and his court against intruders. Disaster of any kind, natural or man-made, signalled to the king's astrologers that the spirits had lost their power and that it was time to move.[10]

After lengthy consultations with his astrologers, King Mindon chose an auspicious site for the new city at the foot of Mandalay Hill on the east bank of the Irrawaddy River north of Amarapura. According to legend, Gautama Buddha had once predicted that

7. The palace at Mandalay. (Vincent, *The Land of the White Elephant*, 1874)

2,400 years after his death, a city devoted to the revival and study of Buddhism would be built there.[11]

In contrast to Ava and Amarapura, Mandalay was situated some distance from the river's banks. Steamships had begun plying the Irrawaddy during King Mindon's reign and he and his ministers worried about a surprise attack by an armed trading vessel. Moreover, the site was also on higher ground and less apt to flood, a decided advantage.[12]

Mindon envisioned a traditional royal city dedicated to Buddhist scholarship, but grander than either of its predecessors, Ava or Amarapura. At the peak of Mandalay Hill, the king constructed a huge gilt statue of Buddha with his arm outstretched protectively over the nascent city. The 'greatest care' was taken to ensure that Mandalay be laid out according to traditional plans.[13] The palace district, for example, was a perfect square with straight white interior roads, 'a conspicuous feature of every Burmese city of which there is any record', according to one reliable source (Plate 7).[14] High walls demarcated the palace's sacred grounds and beyond them a protective moat encircled the royal city. Arches and gilt buildings were features copied from Pegu.

In June 1857, King Mindon and the royal court, accompanied by about 500 members of the clergy, slowly made their way in a splendid procession to the site of the future city. According to a contemporary description, 'the images and Pitakas (Buddhist scriptures) were placed on platforms and carried on the shoulders of men, the images under the shade of eight golden umbrellas, the Pitakas under the shadow of six. The chief Sadaw (monk) had four white umbrellas, and each of the five hundred rahans (monks) two.'[15] A hastily constructed village of temporary palaces and auxiliary buildings sprang up as the king and his court settled down to supervise construction.

The Royal City

King Mindon's plans for Mandalay incorporated many references to Buddhist numerology and religious symbols which reflected the depths of his Buddhist faith. The city's circumference was supposed to be exactly 2400 metres, the year according to the religious calendar and a symbol of the Buddha's prophecy. In reality, however, the walls measured 680 metres from north to south and 650 metres from east to west. They were 12 metres high and over 2 metres thick at the base narrowing to 1 metre at the top. Each wall had three gates. The four centre gates were located at the points of the compass. The parapet of the walls was denticulated and small turrets made of teak with seven tiered roofs stood at equal intervals along the walls. The number seven symbolized the seven paths to enlightenment. At each corner was a fortified tower made of teak and crowned by a seven-tiered tower. A thick earthen wall reinforced the inner side of the wall. A road ran along the outside of the walls and just beyond that was a moat 52 metres wide and 5 metres deep.[16]

Four bridges, each 8 metres wide, crossed the moat to the centre gates. A fifth, the so-called Spirit Gate Bridge (Plate 8), was used for the removal of corpses and avoided by Burmans as unlucky. Bridges were constructed so that the centre portion could be removed allowing the king to circumambulate the city in a religious ceremony held once each year.[17]

Controversy surrounds the question as to whether a number of the king's subjects were actually buried alive under the city's gates and four corners. Some say that King Mindon broke with the tradition of burying live humans whose spirits would then protect the city because he opposed violence. Instead, they said, he buried large jars of oil and placed guardian statues above them. Others, however, argue that the king allowed himself to be persuaded by his astrologers who insisted that, however distasteful, the ritual was absolutely necessary to protect the new city. It is a fact that the inhabitants of Amarapura were terrified at the news of a new capital and the streets of the city were immediately deserted. The 'guardian ghosts' were chosen with great care according to their age, day of birth, and social status. Among them were prepubescent boys and girls, the boys without tattoos and the girls with unpierced ears.[18] In all, fifty-two people were said to have been buried alive, three under each outer gate, one at each corner, four under the Lion Throne, and others under the palace gates and stockades.[19]

8. The Spirit Gate Bridge, Mandalay. (Singer, *Burmah*, 1993)

The Palace Grounds

The royal city of Mandalay was contained within the four walls of the palace and was divided into outer, middle, and inner palace enclosures. Everywhere there were royal troops in colourful green jackets and red helmets, the hilts of their swords gleaming in the sun. Immediately inside the western gate used by most visitors stood the Sacred Tooth Hall (Plate 9). An impressive building of white brick, this temple was an exact copy of one at Pegu built by King Bayinnaung. Lacking a genuine relic, Mindon settled for a copy of the Buddha's tooth. Opposite the Sacred Tooth Hall was a clock tower, similar to ones found in Thailand, which contained a hollow coconut floating in water. As the shell sank, the hours were counted off by striking a gong during the day and a drum at night.[20] In later years, the outer palace grounds also contained the royal mausoleum which held the remains of King Mindon, two of King Thibaw's children, and other members of the royal family. Near by, King Thibaw built a miniature copy of the monastery where he spent his noviciate.[21]

The palace grounds were pleasant with wide, tree-lined streets which ran parallel to the walls and intersected at right angles. Like streets throughout Mandalay, they were dusty in the dry season and thick with mud during the monsoon. 'A high teak palisade, very massive and compact', formed a middle barrier between the outer and inner walls of the palace. Along a tree-shaded esplanade inside the stockade stood the homes of high court officials.[22] The middle palace grounds also contained elephant sheds, horse stables, carriage houses, a large lawn for sporting events and games, and palace mews, which lined the wall in front of the palace.[23]

A brick wall pierced by four heavily guarded gates protected the inner palace grounds. Westerners reluctantly entered through the Red Postern, described as 'a cramped little gateway, which makes the smallest man involuntarily bend his head as he approaches the golden palace.'[24] Hanging from the Red Postern was the Drum of Justice, the very symbol of royal justice. King Mindon himself had directed that the drum be placed there within earshot of his apartments so that a subject in distress might appeal to him for help by striking it at any hour of the day or night.

9. The Sacred Tooth Hall, Mandalay. (O'Connor, *Mandalay and Other Cities*, 1907; Antiques of the Orient, Singapore)

The palace of the Hlut-daw or Supreme Council (Plate 10), an impressive structure where the king's ministers met daily to conduct business, was adjacent to the Red Postern. It was 'raised some feet off the ground, and perfectly open; the roof, richly carved with flowers and figures, and adorned with the usual flamboyant pinnacles, supported on massive teak pillars, painted vermillion colour at the base, and gilt higher up'.[25] King Mindon generally

10. The palace of the Hlut-daw, Mandalay; engraving from the *Illustrated London News*, 20 March 1886. (Antiques of the Orient, Singapore)

declined to attend meetings in the Hlut-daw palace, preferring to leave matters of state to his advisers.

Despite fierce protestations and after a diplomatic crisis, Westerners were compelled to remove their shoes before ascending the winged staircase into the palace. In stockinged feet, they shuffled along a 'colonade of golden pillars' into the Hall of Audience. The large, rectangular throne room was open to the sunlight and tropical breezes on three sides. Originally built at Amarapura by King Tharrawaddy, the entire palace, complete with its multi-tiered roof, had been carefully dismantled and reconstructed piece by piece on its original pillars at Mandalay.[26]

The Lion Throne (Colour Plate 6) rested on a dais 1.5 metres above floor level at the back of the Great Hall. Made of richly carved wood, it was embellished with gilding, vermilion, and small pieces of glass mosaic. The seat was shaped like a lotus in full bloom to symbolize the king's coming Buddhahood. The throne's base was carved in the shape of an hourglass and covered with glass mosaic. Small niches in the base contained figures of lions.[27] Rising gracefully to a height of 78 metres (256 feet) over the Lion Throne was an open tower, crowned by seven tiers, each of them decorated with gold and glass mosaic. The tower acted as a conduit of wisdom to the throne which marked the centre of the universe (Plate 11).[28]

An arched doorway, similar to an aureole and surrounded by numerous flame-like projections, featured images reflecting Burma's syncretic religious beliefs and served as a perfect backdrop to the Lion Throne. The Hindu god Indra or Thagyamin, Chief God of the *Nat*, occupied the apex of the arch. Others included the Hindu god Brahma, the royal *nat*, the Four Keepers of the World, a peacock (symbol of the sun), and a rabbit (symbol of the moon).[29] The exotic setting was enhanced by its backdrop, a red lacquer wall 'gorgeous with bits of looking-glass, porcelain, fragments of coloured glass, a mosaic of queer odds and ends fastened on with resinous gum'.[30] The decorative figures which surrounded the throne were suggestive of forms dating from antiquity.[31] The throne and its background emphasized the king's role as a Bodhisattva. If Westerners found the glass mosaics and peeling red paint a bit tawdry, the king's subjects never failed to be properly impressed.[32]

11. The centre of the universe, Mandalay. (Singer, *Burmah*, 1993)

Like his predecessors, King Mindon occupied the Lion Throne only three or four times each year during special royal audiences. The first marked the New Year (Thingan) when royal princes and high officials gathered before the king to drink the 'ceremonial water of allegiance'. A second audience was held at the beginning of the Buddhist month of fasting. The third came at the end of that month when 'rulers of princely states, the provincial govenors, the princes and Mandalay officials' gathered to drink the water of allegiance. The last two were held on Kadaw or 'Beg Pardon' days when, in a centuries-old tradition, supplicants sought the king's pardon for their misdeeds. Finally, King Mindon sat upon the Lion Throne to view the ordination of his sons, including the ill-fated Thibaw, as Buddhist novices, a ceremony which signified manhood. Soon the king would name one of them his heir.[33]

Several hours before a royal audience (Colour Plate 7), court officials began to assume their appointed places. It was a colourful scene. Four single-tiered white state umbrellas, indicative of the king's presence, were placed around the throne. The heir apparent took his place on a throne described as a 'kind of a cradle' situated below the Lion Throne. The chief queen sat on a throne to the king's right. Her costume—a floor-length shift with long silk cummerbund, a waist-length jacket, and a silk shawl thrown over the shoulders—was similar to but more richly embroidered than the costumes of the lesser queens and princesses. Ministers arrayed in gold and purple sat directly before the king while lesser officials were relegated to the wings. The men wore long satin or velvet coats which flared at the bottom with trousers underneath. Their caps, shaped curiously like pagodas, were made of silk or velvet, often embroidered with flowers or birds. The scene was a true riot of colour, 'a stirabout of rainbows'.[34]

Rank was bequeathed by the king and privileges enjoyed were mandated by strict sumptuary laws. The white umbrella and silk garments embroidered with gold or silver were reserved for the royal family. Court officials nevertheless found numerous ways in which to display their rank. Trying to outdo each other, they went about Mandalay ostentatiously accompanied by servants who carried huge multi-tiered umbrellas of different colours, some 15 feet tall and 6 feet across. They also wore heavy gold chains. The lowliest courtier wore three; a minister, twelve; and the king, twenty-four. Even the shape of a betel box, and the precious metal and jewels of ear-rings, spittoon, or drinking cup reflected the owner's place in the hierarchical society of Mandalay.[35]

Prior to the king's entrance into the Great Hall, the people sat comfortably chatting, some smoking cheroots, others chewing betel. Suddenly, a bell tinkled and, without a pause, the entire assembly fell face forward. King Mindon entered slowly, making his way along a red and gold carpet to the throne. His robes were even more richly embroidered than those of the highest officials and he wore the 'broad gold shoulder belt of the salwe' and 'a tall tapering helmet of gold and gems'.[36] Physically, King Mindon strongly resembled his Konbaung ancestors.

He has a clear and smooth skin, with a bright black eye which twinkles up into quite a Chinese obliquity when he laughs, and that he does every two minutes; his moustache is good, the throat and jaws very massive, the chest and arms remarkably well developed, and the hands clean and small. The retreating forehead, which marks him as a descendant of Alompra [Alaungpaya], was now very conspicuous, and I never saw this feature before in such singular excess.[37]

Like royal cities, the king's audiences adhered carefully to centuries of traditions.

The audience began with the chaunting [*sic*] of a hymn by the Court Brahmins, clad in white and gold. There followed a recitation by one of the 'Messengers of the Royal Voice', of the King's alms, and then each of the princes and noblemen assembled knelt down in homage and renewed his allegiance to the King, and announced through an officer of the Court the presents they had brought to place before His Majesty. The further business of the day was then transacted according to the humour of the King.[38]

The Great Hall with the Lion Throne was the centre of the royal palace. Seven lesser thrones, each shaped like the Lion Throne but made of a different native wood and less sumptuously decorated, occupied other chambers in the palace. King Mindon received ambassadors in the Goose Throne room, 'gazed upon his white elephants' from the Deer Throne (Plate 12), and conducted daily business seated upon the Conch Throne. The Bee Throne was located in the Glass Palace.[39]

The Glass Palace, the king's apartment, was located directly behind the Great Hall. It was covered inside and out with coloured glass mosaic tiles imported from Venice. Crowned by a five-tiered roof, the Glass Palace housed the king's bedroom, four additional

12. The audience hall of the Deer Throne, Mandalay. (Singer, *Burmah*, 1993)

bedrooms for the chief queen and lesser queens, and a dining hall.[40] Although neither King Mindon nor King Thibaw actually lived there, they both received daily visitors in the Glass Palace. The journey from the winged staircase through the palace to the king's quarters was perilous for shoeless Westerners with tender feet. Whether by accident or design, nails protruded randomly from the plank floors. Even seasoned diplomats were forced to lower their eyes respectfully as they watched for the nusiance nails. 'You cannot gaze about you with any degree of freedom,' one noted realistically, 'when you have a lurking suspicion that the next step will drive a nail into the ball of your big toe.'[41]

Except for daily audiences, both Kings Mindon and Thibaw found the glittery Glass Palace intolerable. King Mindon built more modest and comfortable quarters, the Golden Palace Apartment, where he lived until his death in September 1878. (After that the royal apartment was moved to Mandalay Hill where it became a monastery.) On the other hand, King Thibaw, never one for modesty, built an extravagant Italian-style pavilion complete with luxurious living quarters, a library, a theatre, and formal gardens. The gleaming white buildings with their frescos, arched doorways, Roman columns, and Palladian windows stood in sharp contrast to the native teak and bamboo architecture of the rest of the palace.[42]

A plan of the palace complex reveals the Burmese penchant for symmetry (Figure 5). Whether to enhance the circulation of air or to avoid contact with the tainted earth, the palace stood on a raised platform measuring 900 feet by 500 feet and supported by earthen embankments at either end. Approximately 120 wooden buildings, most of them small and identical in style, occupied the platform. The dwellings, which featured pillared corridors and spired roofs of zinc, were arranged in neat rows.[43] To the east of the Glass Palace were the men's quarters and to the west the women's. A boundary, called the Mayapin, clearly delineated the two, and only one man, the king, ever went into the western part of the palace.

Burma's queens and princesses lived in teak houses arranged in orderly rows behind the Glass Palace. The queen mother occupied several buildings on the south side of the palace while the chief queen lived in the Middle Residence, a group of six buildings, directly behind the Glass Palace. The Queen's Audience Hall, which later became a British club, faced west and was located just inside the West Gate of the palace.[44]

The Treasury stood diagonal to the Glass Palace and contained the state archives which were lost in a fire during the British occupation. Gardens with ponds and fountains, an abundance of natural tropical vegetation, canals, and small rest-houses surrounded the palace.[45]

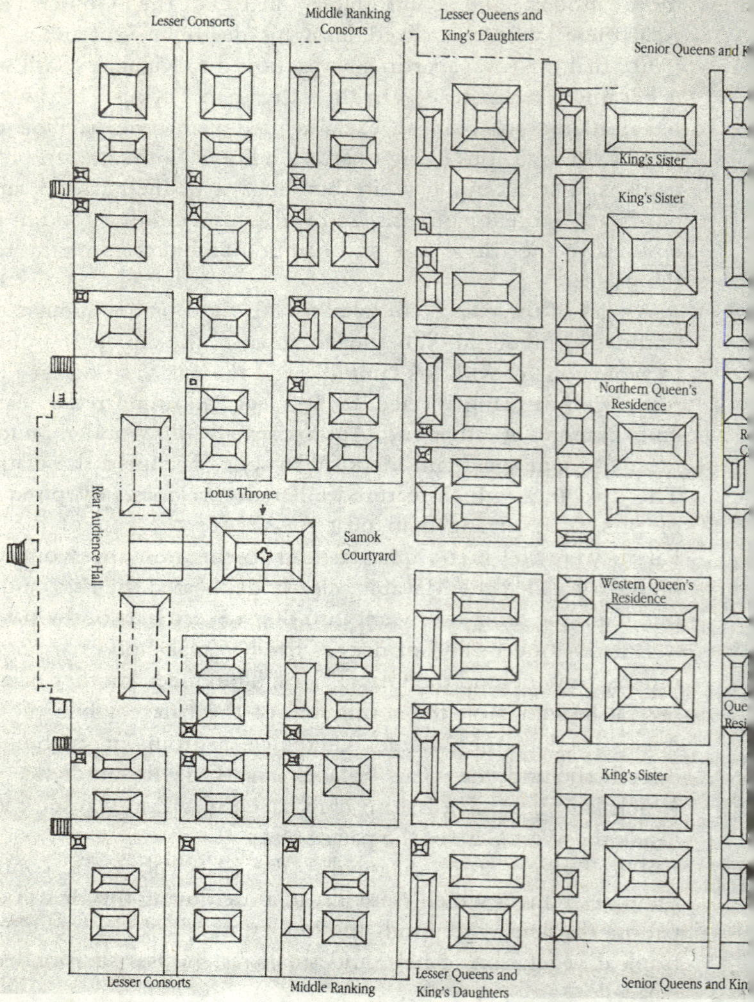

Fig. 5 Plan of the royal palace at Mandalay. (Damrong, *Journey through Burma*, 1991; courtesy of River Books, Bangkok)

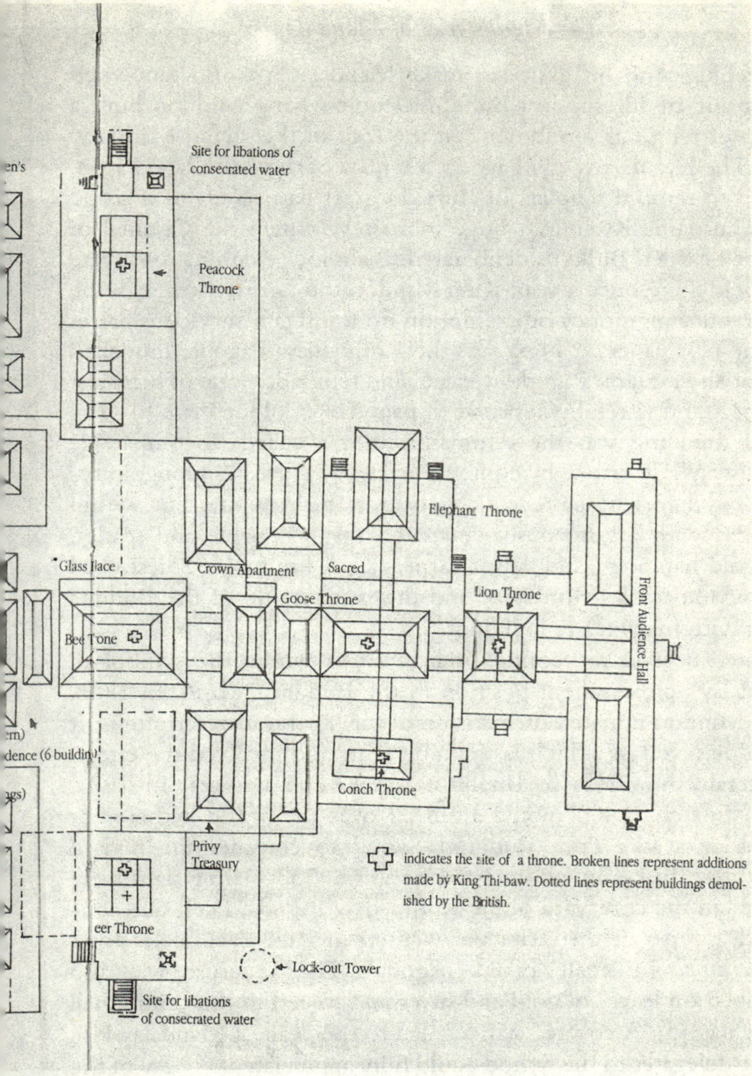
Site for libations of consecrated water
Peacock Throne
Glass Place
Crown Apartment
Sacred
Elephant Throne
Goose Throne
Lion Throne
Front Audience Hall
Bee Tone
Conch Throne
Privy Treasury
eer Throne
Lock-out Tower
Site for libations of consecrated water
indicates the site of a throne. Broken lines represent additions made by King Thi-baw. Dotted lines represent buildings demolished by the British.

The Monasteries of Mandalay

In keeping with his desire to make Mandalay a world-renowned centre for Buddhism and Buddhist studies, King Mindon built a group of religious monuments at the foot of the sacred Mandalay Hill. The Kyauk-taw-gyi Pagoda, a replica of the Ananda at Pagan, served to remind scholars of Burma's great past. Next to it stood the Thu-dama Kyaung, a huge monastery where the Council of Senior Learned Bhikkus deliberated. Religious examinations were also held there once a year. King Mindon took great interest in the exams offering money or exemption from military service as prizes for the best papers.[46] Next was the Kutho-daw Pagoda, modelled on the Shwe-zigon Pagoda at Pagan and reputedly 'one of the most compact and tastefully adorned of pagodas' (Colour Plate 8). The fourth building was the Atumashi, the so-called 'Incomparable Monastery', described by contemporaries as a 'great, oblong, lofty-terraced chapel monastery ... outwardly plain white, but within splendid with a gorgeously decorated shrine, purple and scarlet, and gold hangings, and velvet carpets'.[47] A fire in 1892 destroyed the Incomparable Monastery and the giant statue of the Buddha along with many lesser buildings.[48]

The Kutho-daw became, at least for a while, the symbol of Mandalay's pre-eminent position in the Buddhist world. In 1862, King Mindon initiated the revision of the Theravadan scriptures. It was a massive task which lasted nine years and involved the efforts of literally hundreds of Buddhist monks and novices. In 1871, approximately 2,400 monks from all over the world gathered at Mandalay at the Fifth Buddhist Synod to consider the revised scriptures. To commemorate the occasion, 733 marble slabs were erected in the courtyard of the Kutho-daw Pagoda. Each one contained a verse of the Tripitaka written in Pali. Each stele stood under its own small 'pitaka pagoda'. The scripture was also inscribed on leaves of gold and silver and placed in the central hall of the pagoda. A covered stairway led to the peak of Mandalay Hill where delegates to the Synod could fully appreciate the view of the sacred city.[49]

Other temples and pagodas were located throughout the city. Travellers' accounts never failed to mention the Mya-daung

Kyaung or Queen's Monastery, built by Queen Supayalat, in the southern part of the city (Colour Plate 9). In contrast to King Mindon's tradition-bound temples, Queen Supayalat's monument broke with the past. Its graceful architectural style was reminiscent of Gothic churches in the west and its intricate woodwork paid homage to Burmese craftsmen. The heavily gilded monastery was barely completed when Queen Supayalat and King Thibaw were deposed.[50]

Mandalay's Suburbs

Beyond the royal city lay Mandalay's suburbs flanked by the Irrawaddy River to the west and the Shan Hills to the east. In 1886, the city had a population estimated at 186,000.[51] The city's unpaved streets were distinguished primarily by numbers and were the source of much discomfort. They were 'cut into huge ruts, and the bullock carts went along in the hot weather in a pyramid of dust, and in the rains, a ploughed field would be smooth going by comparison'.[52] Only the Mohier trees whose branches arched across the dusty roads provided relief from the hot sun.

The day began early in Mandalay with the opening of the palace gates at dawn and the endless crowing of roosters. Soon the streets were bustling as housewives in groups of twos and threes chattered on their way to the market where the cacophony grew to a deafening pitch as competing Arabs, Indians, and Chinese hawked their wares (Colour Plate 10). Buddhist monks with shaved heads and orange robes made their way silently through the crowds begging, eyes downcast, rice bowl in hand. The largest of Mandalay's markets was the royal bazaar, the Ze-gyo-daw, located south of the palace. Its narrow alleys thronged with merchants of every race and culture, among them 'the long-robed Persian, the baggy-trousered Shan, the Surati in his cap of cloth of gold, the Chinaman from Yunnan, the Jew from Bussorah and Bagdad, and his rival the expatriated Armenian'.[53] Not far from the main market was the Thibaw Sawbwa, an enclosed camp-site owned by the king and reserved for caravans of Shan tribesmen and others who came to Mandalay to sell their wares.

In a pattern that recalled the ancient villages of Pagan,

Mandalay's neighbourhoods were informally identified by their inhabitants' occupations. Umbrella makers and saddlers lived south of the palace while merchants who specialized in bamboo handicrafts and lacquerware lived to the west. Traders of pottery and miscellaneous items occupied the street leading south to the Arakan Temple. Ironworkers and wheelwrights inhabited the area around 76th Street in the city's north-western section. Along the banks of the canal which wound through the western suburbs a large colony of Indians washed and dried the laundry of the city's wealthy.[54]

Activity in the streets came to a complete halt whenever a court official approached. It was a fact of life which never failed to irk Westerners. 'Before every dignitary rushed along a posse of half-naked lictors, yelling out to the people to bow down, and hitting round about them with their fasces and dahs.'[55] King Mindon's ruling that foreigners could show respect simply by doffing their hats did little to placate outraged Westerners. Loath to give any indication of subservience, they preferred to dart into a side-street and hide at the first sound of an approaching entourage. Burmans, on the other hand, obediently prostrated themselves with their eyes properly lowered—or risked a jail sentence.

The vast majority of Mandalay's citizens lived in humble dwellings constructed of wood planks covered over with roofs of plaited mat. The typical house, whose design had been the same for centuries, stood on posts raised 7 or 8 feet above the ground. Most had only one or two rooms plus a wraparound veranda. Even in the city, land was plentiful and every house had a fenced yard with a small garden. There the family's cart, a huge boat-shaped affair with wooden wheels, was parked. The bullock grazed peacefully near by. Dogs, pigs, and chickens inhabited the space beneath the house and roamed the yard and frequently the neighbourhood at will.[56]

No matter how poor it might be materially, the Burmese home was rich in spiritualism. The precise location of the house was determined by a soothsayer and was important in maintaining the family's well-being. Each house had six supporting posts, all of which were named. The south post, easily identified by the palm leaves which adorned it, was where the family's *nat* spirit resided.[57]

1. A view of old Pagan showing the many pagodas built on the plains surrounding the city. (Kelly, *Burma*, 2nd edn., 1933)

2. Platform of the Shwe-zigon Pagoda, Pagan. (Kelly, *Buṛma*, 2nd edn., 1933)

3. Le Dun or the royal ploughing; from a painting by Saya Chone. (O'Connor, *Mandalay and Other Cities*, 1907; Antiques of the Orient, Singapore)

4. The Ananda Temple, Pagan, King Kyanzittha's greatest achievement. (Kelly, *Burma*, 2nd edn., 1933)

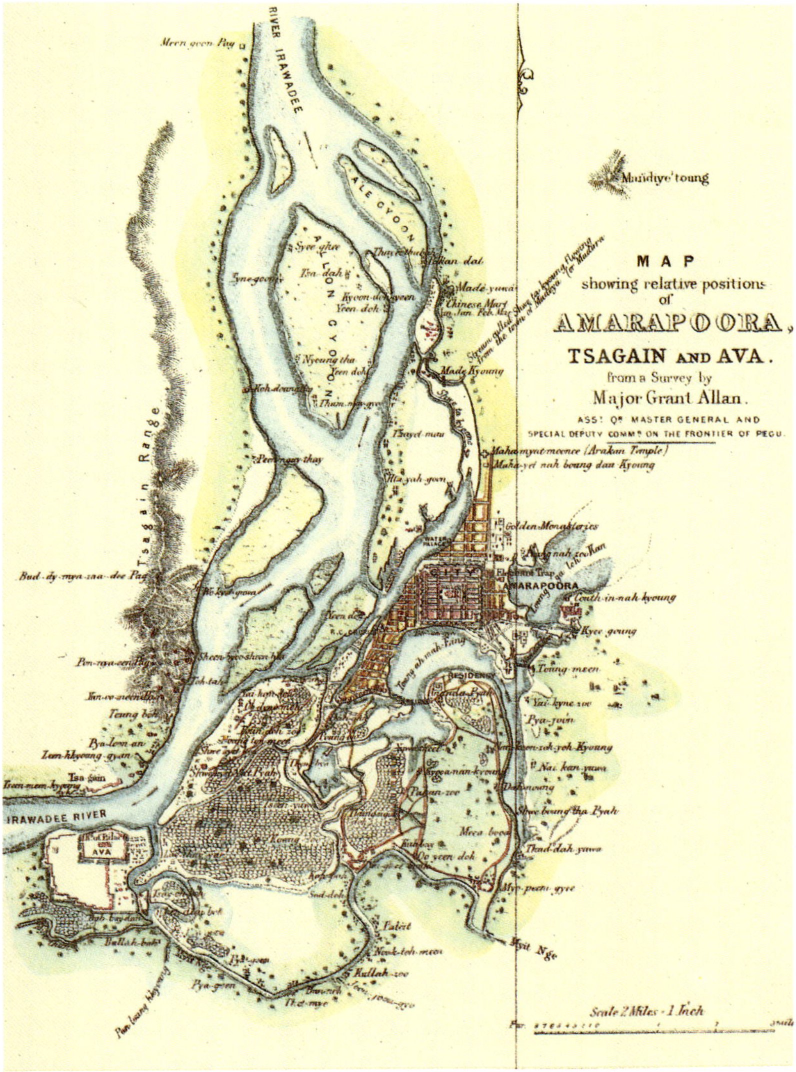

5. Map showing the relative locations of Ava, Amarapura, and Mandalay. (Yule, *Mission to the Court of Ava*, 1857)

6. The Lion Throne, Mandalay, originally built at Amarapura by King Tharrawaddy. (Yule, *Mission to the Court of Ava*, 1857)

7. The royal audience; from a painting by Saya Chone. (O'Connor, *Mandalay and Other Cities*, 1907; Antiques of the Orient, Singapore)

8. The Kutho-daw Pagoda, Mandalay, reputedly 'one of the most compact and tastefully adorned of pagodas'. (Kelly, *Burma*, 2nd edn., 1933)

9. The portico of the Queen's Monastery, Mandalay, built by Queen Supayalat. (Kelly, *Burma*, 2nd edn., 1933)

10. The market-place, Mandalay; hand-coloured engraving from the *Illustrated London News*, 1 May 1886. (Antiques of the Orient, Singapore)

11. King Thibaw's reluctance to leave the palace; from a painting by Saya Chone. (O'Connor, *Mandalay and Other Cities*, 1907; Antiques of the Orient, Singapore)

12. King Thibaw and Queen Supayalat leaving Mandalay; from a painting by Saya Chone. (O'Connor, *Mandalay and Other Cities*, 1907; Antiques of the Orient, Singapore)

13. A British reception in the palace at Mandalay; hand-coloured engraving from the *Illustrated London News*, 3 April 1886. (Antiques of the Orient, Singapore)

14. A shrine on the platform of the Shwe-dagon Pagoda, Rangoon. (Kelly, *Burma*, 2nd edn., 1993)

15. British troops occupy Rangoon during the First Anglo-Burmese War, 1824. (Moore and Marryat, *Birman Empire*, 1825; Antiques of the Orient, Singapore)

16. Merchant Street, Rangoon; hand-coloured engraving from the *Illustrated London News*, 6 March 1869. (Antiques of the Orient, Singapore)

Wealthy Mandalayans built homes of teak with tiled roofs and bamboo or wood floors. Although their homes were more spacious and richly decorated, the basic plan was the same. Many indulged in Western luxuries such as 'chandeliers, bookcases, pictures ... chairs and tables, and occasionally bedsteads'. Even the most common household was apt to own a clock, a kerosene lamp, a mirror, and a 'lounge chair'. Yet despite such luxuries, the typical Burmese house remained drab and dreary. The use of exterior paint was just one more item in a long list reserved for the king's favourites. Brick, however, was reserved exclusively for temples and pagodas. Ever fearful of rebellion, the Konbaung Dynasty made certain that the age-old ban was rigorously enforced. Of course, fire was a constant hazard in Mandalay, particularly in the dry season. Every household kept a bucket and rake handy in the event of a fire in the neighbourhood.[58]

Burmans lived in a world shaped by centuries of tradition while Westerners and Chinese existed beyond the pale of the country's cultural and legal norms. They lived wherever they pleased and used whatever building materials they liked, including brick. A large Chinese community occupied a neighbourhood in the southernmost part of the city, in an area which had once been a northern suburb of Amarapura. When the order came to evacuate the former capital, the Chinese steadfastly refused. In addition to fine brick homes, the immigrant merchants from Yunnan had built a beautiful Chinese temple. Through its circular gate, pedestrians could see the temple's courtyard and its distinctive winged roof. A haughty inscription over the doorway, which few could read for it was in Chinese, said: 'Even in the country of the Barbarians there is Enlightenment.' King Mindon's ministers had urged him to force the Chinese to move but he wisely chose to leave them alone. They were, after all, prosperous merchants, and their trade was important to Upper Burma's economic well-being. In the end, Mandalay gradually expanded until it engulfed the Chinese community.[59]

Many Westerners lived in a district bounded by the south-west gate of the palace, the so-called 'accursed gate', and the Irrawaddy River. Others, mostly traders, lived in bungalows scattered throughout the city. There was also a large contingent of Italian and French army officers who lived in quarters throughout the city and were employed by the king to instruct and improve the skills of the native soldiers. The mercenaries, a general or a colonel every one, went without pay for long periods of time.[60]

In general, the kindly King Mindon welcomed expatriates to Mandalay. He tolerantly allowed foreign clergy to establish their churches. There was a Roman Catholic Cathedral in the city's south end, not far from the royal bazaar. Italian and French priests were active in the city, admonishing their countrymen not to go astray in this exotic, godless land. King Mindon commissioned the building of almshouses and small hospitals where Western medicine was practised.[61]

Undoubtedly, the 'most notable foreigner' in Mandalay was Dr Marks, an Anglican priest, who established a church and boys' school (Plate 13) in 1868 on behalf of the Society for the Propagation of the Gospel. Always interested in education, King Mindon donated land for the project and even enrolled his sons, including Thibaw, in Dr Marks's classes. For a time, the king and the minister were friendly, meeting frequently to converse at length on topics of mutual interest, particularly Buddhism. The two had a falling out, however, when Dr Marks refused to appeal to London for the return of the port of Bassein to Upper Burma. Court officials soon let the Anglican priest know that if he valued his life it would be best to leave Mandalay.[62]

King Mindon was also interested in economic reforms and looked to the West for ways in which to improve the lot of his small land-locked kingdom. Industrialization seemed to offer compensation for the loss of revenue from Lower Burma. The king imported both men and machinery from Europe and began to produce lac, cutch, sugar, cotton, and silk. He also had a passion for making guns to defend his small kingdom. But the economy of Upper Burma remained stagnant and the king was usually in arrears when it came to meeting the payroll for his foreign managers.[63]

If the economy of Upper Burma improved at all, it was because King Mindon significantly increased trade with the British in Lower Burma through a series of treaties. He built a fleet of steamships which plied the Irrawaddy and strung up telegraph lines to

13. Dr Marks's school, Mandalay. (Singer, *Burmah*, 1993)

improve communications. He also introduced money into Upper Burma's barter economy. However, the coins were usually in short supply and were suspect because the metals were frequently mixed with alloys. In addition, the value was not fixed by law. To ensure a regular income for the government, King Mindon levied taxes on households and paid officials set salaries in place of the earlier feudal arrangements whereby their livelihood was maintained by villages of peasants. By all accounts, the king made a determined effort to improve the material lot of his subjects and to overcome the xenophobia of his predecessors. Unfortunately, when King Mindon died in 1878, his economic programme had not gone far enough to bring about real progress in Upper Burma.[64]

King Thibaw's Reign, 1878–1885

King Mindon's death plunged the royal city into confusion and intrigue as plots and counterplots were played out behind the palace walls. Only a year earlier, the crown prince had been murdered and the bereaved old man refused to appoint another to avoid increasing dissension and hostility in his household. Finally, he suggested dividing the country between his three older sons, a compromise which satisfied no one. Following his death, King Mindon's ministers conspired to make the least talented younger son, Thibaw, a puppet king. They failed, however, to take into account his wife, Queen Supayalat, who was every bit as cunning and ambitious as they and who had complete control over her husband. It was Queen Supayalat who planned the cold-blooded massacre of eighty-six royal opponents in February 1879. The massacre triggered official outrage and protests from the British, and there were even rumours of an invasion from Lower Burma. British residents in Mandalay, including the Resident, hastily packed up and headed south. To the court's relief, the British government was otherwise engaged fighting wars in Afghanistan and South Africa and declined to intervene in Upper Burma.[65]

Aside from foreign condemnation, King Thibaw faced a constant stream of domestic problems, resulting from Mandalay's loss of control over the countryside. The ever-rebellious Shans, Kachins, and other ethnic groups took the opportunity to rise up against the untested king. Law and order broke down further as bandits and renegades terrorized Upper Burma. The hapless king turned to his wife for guidance, and Queen Supayalat unhesitatingly ordered military troops to take all necessary measures to quell the rebels. The country finally settled into an uneasy peace when King Thibaw's popular chief rival for the throne died in 1885. The British in Lower Burma continued to keep a wary eye on the north.[66]

King Thibaw proved to be such a striking contrast to his predecessor that it was openly suggested that perhaps King Mindon was not, in fact, his father. What was certain was that his mother was Shan, which for many explained his lack of character and self-discipline. King Thibaw's former teachers described him as 'surly and morose', particularly when things did not go his way on the cricket field. On more than one occasion during his schooldays, the young Thibaw had only barely escaped a sound whipping.[67]

There could be no denying that the bloody massacre which inaugurated his reign had left deep scars. Nightmares tormented him with vivid scenes of palace guards clubbing his brothers, sisters, nieces, and nephews into unconsciousness or death. The bodies were thrown, some still half-alive, into a huge pit. The grave was covered over and trampled down by the executioners. Horribly, after a day or two, the mound of earth began to rise up. This time the royal elephants were brought in to complete the job. Understandably, the king became nervous and easily frightened. It was then that he began drinking heavily and was frequently seen with his face blotched and bloated.[68] Fearful of assassination, King Thibaw constructed a watch-tower close to the royal apartments and posted a guard there day and night.

Notwithstanding questions about his paternity, King Thibaw strongly resembled the Konbaungs with his high, sloping forehead and flat nose. In his personal habits he was careless and even slovenly. Frequently, his robes were wrinkled and soiled and the topknot wig (which covered the former monk's bald head) was dishevelled and crooked. Although he often appeared bored in private meetings, the king obviously enjoyed public audiences. Then he was affable and animated, asking questions and reciting moral precepts to impress the crowd.[69]

As to Queen Supayalat's ancestry, there could be no doubt. She was descended from the Konbaungs on both sides. Handsome and regal in her bearing, she too had the characteristic 'Alompra forehead'. Queen Supayalat was insanely jealous and had a temper which was 'passionate and resentful to the last degree', traits also reminiscent of her forebears.[70] Driven by jealousy, she demanded to be the only queen and King Thibaw meekly agreed to abandon the tradition of having four queens to ensure a male heir. Rumours abounded of the cruel deaths suffered by comely maidens upon whom the king's gaze might have lingered a bit too long. The couple had only two children, a son and a daughter who both died in the smallpox epidemic of 1880. In her jealousy, the queen thus deprived Upper Burma of an heir to the throne.[71]

One disaster after another struck Mandalay in 1880. In January, a smallpox epidemic broke out striking rich and poor alike. King Thibaw's infant son and daughter were among those who died. Shortly after that, a magnificent ruby, one of the Crown jewels, disappeared. Most ominous of all, a tiger escaped from his cage in the royal gardens and mauled a worker to death before being recaptured. Obviously, the royal *nat* were offended. Astrologers urged King Thibaw to abandon Mandalay, but, unlike earlier capitals, Mandalay had many substantial buildings, including King Mindon's factories, owned by the Crown. The only remaining solution was to propitiate the gods with human sacrifices. King Thibaw agreed and the chief astrologer calculated that at least 600 human lives, including equal numbers of men, women, boys, girls, soldiers, and foreigners, would have to be sacrificed. Palace guards began rounding up the would-be victims. More than 100 people were arrested. Steamers headed south were 'crowded to suffocation' with hysterical Burmans. Only British outrage put a stop to this latest Konbaung insanity.[72]

Seeing that their advice went unheeded, one more omen came as no surprise to the astrologers. In 1884, the Arakan Temple south of the city burned to the ground. That the precious Maha-muni image of the Buddha inside was miraculously saved did little to pacify the soothsayers' hysteria. (The image had been cast in AD 146 by the king of Arakan and was said to have miraculous powers.) The temple attracted thousands of the faithful who believed that the image's protective powers extended over all of Burma. The fire, his critics pointed out, was just one more indication that King Thibaw had fallen out of divine favour, if indeed he had ever enjoyed it in the first place. The king hastily started construction on another temple, but Upper Burma fell to the British before its completion. The astrologers' worst fears had become reality.[73]

Shortly after King Thibaw ascended the throne, one Western diplomat had grimly predicted that his reign would be short and that he would not die a king. King Thibaw ruled Upper Burma for only seven years. Indecisiveness and passivity characterized his reign except for outbursts of bellicosity against the British. In an effort to increase his popularity, he strongly hinted at a forthcoming invasion of Lower Burma. The king's ill-conceived threats caused the British to reinforce their frontiers. Trade between Upper and Lower Burma fell off drastically, and merchants in Rangoon urged their government to intervene.[74]

King Thibaw's dalliance with the French and the cancellation of a British trade monopoly finally led to a total breakdown in relations with the British. The king found a much-needed ally in the new French Consul who arrived at court in May 1885. By midsummer, Burma had signed a treaty allowing the French, among other things, to establish a bank in Mandalay and to build a railway from the capital to Toungoo in Lower Burma. Rumours reached Rangoon that King Thibaw intended to grant the French control of the royal monopolies. In August, India's Govenor-General, Lord Dufferin, threatened King Thibaw with war if he permitted France to displace the British in the economy of Upper Burma. The situation turned into a crisis when a secret letter revealed that, in return for the concessions, the French would supply King Thibaw with arms and other war supplies but only after they were able to put down a troublesome uprising in Tongking. Confronted by the British, the Foreign Ministry in Paris decided that it was best to withdraw altogether from Burma. The French Consul left Mandalay, but the Foreign Minister in Paris stubbornly refused to acknowledge that Burma was, in fact, within Britain's sphere of influence in South-East Asia. Before the two world powers had settled the issue, King Thibaw further angered the British by allowing the Hlut-daw to cancel the Bombay Burmah Trading Company's long-held monopoly in the Ningyan teak forests. Outraged, the British prepared for war.[75]

In reality, contingency plans for an invasion of Upper Burma had been made months earlier. The British issued an ultimatum on 30 October, but it met with a bellicose reply from the poorly advised King Thibaw. The invasion began on 14 November 1885 and the Third Anglo-Burmese War was over in two weeks.[76]

On the morning of 28 November, the British fleet commanded by Colonel Edward Sladen docked in Mandalay. From a watch-tower, Queen Supayalat had observed the British ships approaching. Hysterically, she admitted what others had dared only to think. 'It is I—I alone—I, the queen that have brought destruction to the king my husband and my people. It is I—I alone.'[77] Then she regained her composure and tried to reason with her husband who was frantically proceeding with preparations to escape to the north. Queen Supayalat was several months pregnant and simply could not withstand the trip. But King Thibaw refused to listen as servants hastily assembled more than 50 elephants and 200 ponies to carry the court and their belongings.

Shortly after docking, Sladen dispatched a letter to the chief minister ordering him to bring the king to Sladen's ship, the *Doowoon*, at noon. When the king failed to appear, troops, accompanied by a military band, went ashore to arrest the king. Large crowds gathered along the way and watched quietly and passively as the British marched through Mandalay.[78]

At about 3 p.m., Colonel Sladen himself arrived at the palace grounds. Contemptuously, he strode into the palace without removing his shoes. King Thibaw, reportedly in tears, begged that he and Queen Supayalat be allowed to remain in the palace until after the birth of the baby. Sladen refused, saying that the couple should prepare to leave Upper Burma immediately. Then he relented somewhat and agreed to allow the king and queen to spend one last night in the palace. The British commander promised that no troops would enter the palace but would camp outside. In return, he demanded that the king's ministers guarantee with their very lives that King Thibaw not escape.[79]

Following a restless night in the Hlut-daw, Sladen found the king and queen terrified and hiding in the royal apartments. Their servants had fled during the night, and a crowd of peasant women, several hundred strong, had forced their way into the palace, demanding jewellery, silk, and other items of value. British soldiers appeared in time to save the king and queen, but Sladen realized that assassination was a very real possibility and assigned twenty-five soldiers to guard the couple.[80]

On 29 November, King Thibaw and Queen Supayalat left Mandalay forever. The king had been outside the confines of the palace grounds only three or four times in seven years and had never been farther than 5 miles from Mandalay. Terrified of what was ahead, he resisted leaving (Colour Plate 11). 'Finally, Sladen and two staff officers walked up the several steps of the pavilion. They stood over Thebaw [*sic*], who had never in his life had any one stand over him in such a manner. Thebaw stood up, and they urged him down the steps.'[81]

Slowly, King Thibaw gathered his courage. He insisted on walking from the palace to the river bank because no suitable carriages were available. After walking for several miles, the weary king and queen gave in and rode the rest of the way through the streets of Mandalay in an ordinary bullock cart (Colour Plate 12).[82]

> The procession became lost and went at least two miles out of its way. It was led by two regiments of Indian infantrymen, a screwgun battery, and then the King and Queen. There were eight white umbrellas—one less than usual—over the King and Queen, which was a most subtle way of telling the populace that Thebaw was no longer King. They were flanked by double-rows of fixed-bayonets, regimental colours, the clash of bands, and the Hampshire Regiment followed in the rear.[83]

The people of Mandalay respectfully prostrated themselves as the procession passed by, but a great cry of lamentation rose from the usually silent crowd at the sight of their prisoner king. The British military band unsuccessfully attempted to drown out the pitiful sound. When they arrived at the dock, King Thibaw hesitated. Finally, Queen Supayalat took him by the hand 'as a mother leads her child when it is lost and afraid', and the pair walked up the gangplank.[84]

The king and queen were exiled to the town of Ratnagiri on

the west coast of India. They were free to move about the town and received an annual pension of 100,000 rupees a year, a generous allowance. Although a large contingent of Burmans initially followed the royal couple into exile, their numbers gradually dwindled to seven or eight. Increasingly lonely and isolated from his countrymen, King Thibaw remained in India for thirty-one years until his death at fifty-eight in 1916. After that, Queen Supayalat was allowed to return to Rangoon where she lived until her own death in 1925 at the age of sixty-eight.[85]

The British Occupation

Within twenty-four hours of the king and queen's departure from Mandalay, the royal palace was looted. Anything of value that could be carried off disappeared. The Burmans blamed Colonel Sladen, and he, in turn, accused the king's ministers of stripping the palace clean. An investigation of the charges and counter-charges came to nothing.[86]

In the following weeks, the palace was taken over by the British (Plate 14). The Queen's Audience Hall became the Upper Burma Club, frequented by military officers and high-ranking civil servants (Colour Plate 13). The Great Hall, where King Mindon once presided, was converted into an Anglican Church. Lesser rooms became offices, dining-halls, and meeting-rooms. Furniture was moved out of the way or discarded and royal thrones, including the Lion Throne, were lost in the process. Troop barracks were constructed inside the palace walls. Written some years later, the following nostalgic description neatly captures the imperial attitude:

> Evening after evening the women come here a little pale, as English women are in the East: evening after evening the men gather from the polo-ground and the tennis court; the band plays, the markers call the scores at the billiard tables in the Queen's Hall of Audience. On the mirrored walls hang the latest telegrams from Europe; at the foot of the throne, picture papers portray the incidents of some city pageant, some royal procession, some battle of the Empire. Long men lie in the easy chairs under swinging fans or sit in the Queen's inner chamber playing at whist and poker. The swish of soda-water bottles, the crack of ice, the click of billiard balls; such [was] the western end of the palace.[87]

14. Burmese ladies taking afternoon tea with Lady Dufferin in the palace at Mandalay; engraving from the *Illustrated London News*, 3 April 1886. (Antiques of the Orient, Singapore)

For fifteen years, the palace served as the Upper Burma Club. In 1901, the Viceroy of India, Lord Curzon, visited Mandalay. Unlike many of his colleagues, Curzon appreciated Burma's culture and realized the importance of preserving the palace. Curzon ordered that the club and church be moved and that the central part of the palace be preserved—not, he noted, because of its historical importance but 'from its value as a model—the only one that will before long survive—of the civil and ceremonial architecture of the Burman Kings'.[88] Preservation, not complete restoration, was what Curzon had in mind. Certain buildings, chosen for their 'character, condition, and artistic merit' were to be maintained. This included making necessary repairs, particularly of carved woodwork and occasional cleaning. Regilding, which Curzon did not care for, was not to be done except in 'cases of obvious necessity' because it was expensive and it resulted in a 'sharp and unnecessary contrast' between buildings.

Curzon was well aware that the progressive decay of the wooden palace was inevitable. 'It will be for some successor of mine,' he wrote, 'to decide whether, as the structure grows older and gradually moulders into decay, it will be worth while attempting renovation on a larger scale than I have foreshadowed.'[89]

Other structures scheduled for preservation included the 'fanciful tiered wooden pavilions over the gateways of the old City Wall', and the bridges crossing the moat. 'It might be worth while to give a little money to the repair of the Queen's Monastery,' he noted, adding that if expensive regilding were required, it would be up to the Buddhists to do it themselves.[90]

At the same time the Viceroy began construction of a residence for the Lieutenant-Govenor of Burma inside the northern wall of the palace grounds. Curiously, the building's architectural style was Burmese, not British. Perhaps Curzon was merely being consistent. Or possibly he was subtly reminding the natives of the reality of their present situation, that the Lieutenant-Govenor ruled in the name of the Queen of the British Empire.[91]

With the British occupation, the rest of Mandalay was left much as it was except for the introduction of traffic signs. A racecourse, as well as a rifle-range, was laid out west of Mandalay Hill. Laws forbidding the use of brick in construction were repealed and new brick buildings appeared throughout the city.[92]

By the 1930s, visitors to Mandalay found well-maintained palace walls and spired towers and a moat filled with clean water. Unfortunately, as Curzon had predicted, time had taken its toll on the teak buildings. Only the Throne Hall, the king's apartment, the royal tombs, the clock tower, and the Sacred Tooth Hall remained as relics of Burma's royal past. The once sacred ground was now called Fort Dufferin and featured parade gounds, married soldiers' quarters, and separate hospitals for males and females. Visitors could view a model built by British craftsmen to get an idea of what the palace grounds once looked like.[93]

Like all of Burma, Mandalay suffered terrible devastation during World War II. The royal palace ended its days as the headquarters of the Japanese army. With independence in 1947, the city recovered and settled into a routine based on the traditional pattern of Burmese life. Fort Dufferin became the headquarters of Burma's military. Officials took little interest in preserving royal relics. Deterioration of the remaining royal buildings continued until only the platform on which the king's apartment once stood remained. Once and for all, the monarchy was relegated to the past and the seat of political and economic power shifted to Rangoon. The city, however, retained its influence as the country's cultural capital. Mandalay continues today to represent the very essence of the culture and history of Burma. It is located at the foot of a mountain sacred to Buddhism, the religion of the vast majority of the people. It sits on the banks of the Irrawaddy River, the ancient waterway which served to unite the country. And it dominates the very heartland of Burma in the neighbourhood of Pagan, Ava, and Amarapura. Perhaps more than Mindon ever expected, his marvellous Mandalay lives on in the minds and hearts of his countrymen nearly a century and a half after its founding.[94]

1. On the political history of the origins of Mandalay, see D. G. E. Hall, *A History of South-East Asia*, 4th edn., London: Macmillan Education Ltd., 1981, pp. 659–71. Notable works on Mandalay include V. C. Scott O'Connor, *Mandalay and Other Cities of the Past in Burma*, London: Hutchinson and Co., 1907; reprint edn., Bangkok: White Lotus Co., 1987; Paul Strachan, *Mandalay: Travels from the Golden City*, Gartmore: Kiscadale Publications, 1994; E. C. V. Foucar, *Mandalay the Golden*, London: Dennis Dobson, 1963. See also HRH Prince Damrong Rajanubhab, *Journey through Burma in 1936*, trans. Kennon Breazeale, Bangkok: River Books, 1991, p. 48.

2. O'Connor, *Mandalay and Other Cities*, p. 4; Damrong, *Journey through Burma*, pp. 78–9.

3. Hall, *History of South-East Asia*, p. 426.

4. Ibid., p. 173.

5. Ibid., p. 625; O'Connor, *Mandalay and Other Cities*, pp. 372–4; Damrong, *Journey through Burma*, p. 48.

6. O'Connor, *Mandalay and Other Cities*, pp. 128–9.

7. Ibid., pp. 119–20.

8. Henry Yule, *A Narrative of the Mission to the Court of Ava in 1855*, Calcutta: Thacker, Spink and Co., 1857, p. 131.

9. Ibid., p. 31.

10. Shway Yoe [Sir George Scott], *The Burman: His Life and Notions*, 1882; reprint edn., Gartmore: Kiscadale Publications, 1989, pp. 481–2.

11. O'Connor, *Mandalay and Other Cities*, p. 4; Damrong, *Journey through Burma*, pp. 78–9.

12. O'Connor, *Mandalay and Other Cities*, p. 4; Damrong, *Journey through Burma*, pp. 78–9.

13. O'Connor, *Mandalay and Other Cities*, p. 6; Damrong, in *Journey through Burma*, p. 84, makes the argument that Mandalay's plans were based on those of Pegu.

14. O'Connor, *Mandalay and Other Cities*, p. 62.

15. Ibid., pp. 6–8; Damrong, *Journey through Burma*, p. 80.

16. Damrong, *Journey through Burma*, pp. 79–84.

17. Ibid., p. 79.

18. O'Connor, *Mandalay and Other Cities*, p. 6; Foucar, *Mandalay the Golden*, p. 27.

19. Shway Yoe, *The Burman*, p. 482.

20. Damrong, *Journey through Burma*, pp. 49, 85; Albert Fytche, *Burma Past and Present*, 2 vols., London: C. Kegan Paul and Co., 1878, Vol. 1, p. 252.

21. Damrong, *Journey through Burma*, pp. 88–9.

22. Ibid. Descriptions of this part of the palace grounds are confusing. According to Fytche (*Burma Past and Present*, Vol. 1, p. 252), the teak palisade was the outermost wall. Shway Yoe (*The Burman*, p. 478) agrees with Damrong's account.

23. Damrong, *Journey through Burma*, p. 89.

24. Shway Yoe, *The Burman*, p. 478; Foucar, *Mandalay the Golden*, p. 49.

25. Shway Yoe, *The Burman*, p. 479; O'Connor, *Mandalay and Other Cities*, p. 34.

26. Shway Yoe, *The Burman*, p. 473; O'Connor, *Mandalay and Other Cities*, pp. 82–6.

27. Damrong, *Journey through Burma*, p. 94.

28. Ibid., p. 93.

29. Ibid., p. 94.

30. Shway Yoe, *The Burman*, p. 473.

31. O'Connor, *Mandalay and Other Cities*, pp. 88–90.

32. Shway Yoe, *The Burman*, p. 479.

33. Ibid., p. 471; Damrong, *Journey through Burma*, p. 95.

34. Shway Yoe, *The Burman*, p. 474.

35. Ibid., pp. 406–12.

36. Ibid., pp. 474–5; Foucar, *Mandalay the Golden*, p. 38.

37. O'Connor, *Mandalay and Other Cities*, p. 94.

38. Ibid., pp. 86–7.

39. Ibid., pp. 94–8; Damrong, *Journey through Burma*, pp. 96–7.

40. Damrong, *Journey through Burma*, p. 98.

41. Shway Yoe, *The Burman*, p. 462.

42. O'Connor, *Mandalay and Other Cities*, p. 96; Damrong, *Journey through Burma*, p. 99.

43. O'Connor, *Mandalay and Other Cities*, pp. 82–4.

44. Damrong, *Journey through Burma*, pp. 100–1.

45. Ibid., p. 89.

46. Ibid., p. 108.

47. Shway Yoe, *The Burman*, p. 172.

48. O'Connor, *Mandalay and Other Cities*, pp. 48–60; Damrong, *Journey through Burma*, 107–10.

49. Hall, *History of South-East Asia*, pp. 659–67.

50. O'Connor, *Mandalay and Other Cities*, p. 80; Damrong, *Journey through Burma*, p. 111.

51. J. G. Scott, *Burma from the Earliest Times to the Present Day*, London: T. Fisher Unwin Ltd., 1924, p. 362. Scott also wrote under the pen-name Shway Yoe.

52. Shway Yoe, *The Burman*, p. 541; Foucar, *Mandalay the Golden*, p. 44.

53. O'Connor, *Mandalay and Other Cities*, pp. 66–78; Foucar, *Mandalay the Golden*, p. 46.

54. Shway Yoe, *The Burman*, p. 541; O'Connor, *Mandalay and Other Cities*, pp. 66–70.

55. Shway Yoe, *The Burman*, pp. 472–3, 543.

56. Fytche, *Burma Past and Present*, Vol. 2, p. 66.

57. Shway Yoe, *The Burman*, pp. 75–80.

58. Ibid.

59. Ibid., pp. 540–1; O'Connor, *Mandalay and Other Cities*, pp. 79–80.

60. Foucar, *Mandalay the Golden*, pp. 44–8.

61. Ibid., pp. 47–53; O'Connor, *Mandalay and Other Cities*, p. 76.

62. Shway Yoe, *The Burman*, p. 468; Foucar, *Mandalay the Golden*, pp. 48–53.

63. Foucar, *Mandalay the Golden*, pp. 46–7.

64. Ibid.; Hall, *History of South-East Asia*, p. 671.

65. Hall, *History of South-East Asia*, pp. 671–3.

66. Ibid., pp. 676–7.

67. Foucar, *Mandalay the Golden*, p. 51.

68. Shway Yoe, *The Burman*, pp. 466–7; Damrong, *Journey through Burma*, p. 100; O'Connor, *Mandalay and Other Cities*, pp. 20–2.

69. Shway Yoe, *The Burman*, pp. 467, 475–6.

70. O'Connor, *Mandalay and Other Cities*, pp. 27–8.

71. Shway Yoe, *The Burman*, p. 457.

72. Ibid., pp. 482–4.

73. Damrong, *Journey through Burma*, pp. 112–13.

74. Shway Yoe, *The Burman*, pp. 470–6.

75. Hall, *History of South-East Asia*, pp. 677–80.

76. Ibid., p. 680.

77. Charles Lee Keeton III, *King Thibaw and the Ecological Rape of Burma*, Delhi: Manohar Book Service, 1974, p. 286.

78. Ibid. For a sympathetic account of Supayalat's reign, see H. Fielding, *Thibaw's Queen*, London: Harper and Brothers, 1899. See also Scott, *Burma*, pp. 327–30.

79. Keeton, *King Thibaw*, pp. 286–7. Previously, during Mindon's reign, Sladen had been British Resident in Mandalay. In a conversation with Fytche, Commissioner of Lower Burma, Mindon had noted that Sladen was 'rare even among foreigners.... He works for the interests of both countries.' See Fytche, *Burma Past and Present*, Vol. 2, p. 270.

80. Ibid., p. 287.

81. Keeton, *King Thibaw*, p. 288; Fielding, *Thibaw's Queen*, pp. 291–4.

82. Keeton, *King Thibaw*, p. 289; Fielding, *Thibaw's Queen*, pp. 291–4.

83. Keeton, *King Thibaw*, p. 289.

84. Ibid., 290; Foucar, *Mandalay the Golden*, pp. 206–25.

85. Damrong, *Journey through Burma*, p. 32.

86. Ibid., pp. 81–2. For an account of the British disorganization in taking over Mandalay, see Scott, *Burma*, pp. 331–3.

87. O'Connor, *Mandalay and Other Cities*, pp. 72–3.

88. Ibid., pp. 417–19.

89. Ibid., p. 419.

90. Ibid., p. 422.

91. Damrong, *Journey through Burma*, pp. 82–3.

92. Ibid., p. 74.

93. Ibid., pp. 83–101.

94. Strachan, *Mandalay*, pp. 35–48. Strachan offers a realistic account of an energetic and intense city. Much of the centre of Mandalay was destroyed in a fire in 1980. Since then the city has undergone a modernization programme.

4
Rangoon: City of Conquerors

RANGOON, Burma's seat of government today and the third in this trilogy of cities, achieved fame and fortune as a capital of the British Empire. Founded in 1755 by King Alaungpaya to celebrate the extension of his authority over the Mon delta lands of the south, Yangon, as it was named, quickly settled into obscurity after a brief moment of fame. The fact was that the xenophobic Konbaungs cared little for contact with the outside world, so the port of Rangoon languished for a century until the British captured it in the Second Anglo-Burmese War. The British greatly admired King Alaungpaya, who unified Burma against all odds, and no one more so than the eighteenth-century diplomat Michael Symes who viewed him as a demi-god. 'His firmness emancipated a whole nation from servitude,' he wrote, 'and, inspired by his bravery, the oppressed in their turn subdued their oppressors.'[1]

Symes became the first British official to visit Rangoon when he stopped there on his way to Ava in 1795 on a mission from the Govenor-General of India.[2] Rangoon had a population of approximately 30,000 and stretched for nearly a mile along the river's east bank and inland about a third of a mile. The diplomat was particularly impressed with the commercial potential of the river town (Figure 6). Symes described three wharfs with cranes which extended out over the river and two 'commodious wooden houses' near by where traders and merchants met daily. In detail he told of a 'spacious' brick customs house with a sturdy tile roof and the merchants who met there. He wrote:

> Here are to be met fugitives from all countries of the East, and of all complexions; the exchange, if I may so call the common place of their meeting, exhibits a motley assemblage of merchants, such as few towns of much greater magnitude can produce: Malabars, Moguls, Persians, Armenians, Portuguese, French, and English, all mingle here and are engaged in various branches of commerce.

Early in his visit Symes realized that Rangoon was 'one of the

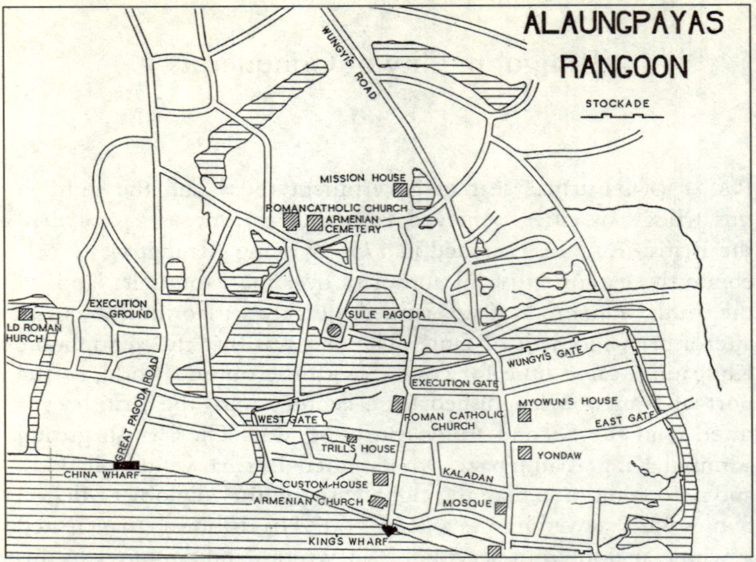

Fig. 6 Map of Alaungpaya's Rangoon. (Pearn, *History of Rangoon*, 1939)

world's finest seaports', except during the monsoon season when seasoned pilots wisely headed for Negrais, Martaban, or Mergui. His enthusiasm was understandable given the fact that there was not a single deep water port in all of the western Bay of Bengal. Symes's imagined commercial plans easily extended beyond Rangoon to timber-rich Pegu and north to Ava and even China where he envisioned huge markets for British goods.[3] It should come as no surprise that Symes's interest in Rangoon's commercial prospects aroused suspicion on the part of government officials. He managed, however, to dampen his enthusiasm long enough to persuade the king to permit the posting in 1796 of Captain Hiram Cox to Rangoon as the first British Resident in Burma.[4]

Symes's Rangoon was an unimpressive wooden stockade surrounded by a moat which he described as an 'indifferent fosse'. Wooden bridges led to the town's five gates. A battery of twelve cannons, in 'awful' condition, were trained on the river and the 'considerable' town on the opposite side. The north wall of the fort had two gates; each of the others had only one. Although the streets inside the fort were narrow, they were well swept and had channels down the middle to drain off the rainwater and wooden planks for pedestrians. The fort housed only government officials, wealthy merchants, and other 'persons of consideration'. Symes managed to rent two houses there for himself and his party. Men of 'inferior rank' lived outside the walls in the so-called 'suburbs' of Rangoon. An area called Tackally was reserved for prostitutes.[5]

Rangoon looked much like the other towns and villages in eighteenth-century Burma. Inhabitants lived in simple, unpainted wooden houses raised several feet off the ground and covered over with roofs of palm. Side walls of thatch or light wood could be raised during the day for ventilation. Earthen pots filled with water sat on every roof and a bamboo handled rake was kept handy to fight frequent fires. Men found guilty of crimes such as theft were sentenced to patrol the streets at night on the look-out for fires.[6]

Symes liked the Burmans, whom he described as 'a lively, inquisitive race, active, irascible, and impatient'. The women, he wrote, were 'inclined to corpulence' and more closely resembled the Chinese while the men were short but 'active and athletic'. A well-travelled man, Symes declared that he had never seen so many strange practices among a single group of people. For example, Burmese men tattooed their entire bodies with fantastic shapes in an attempt to ward off evil spirits. The women stained the palms of their hands and finger-nails with henna and covered their faces and breasts with sandalwood powder. Girls were taught from a young age to rotate their elbows outward in such a way to make them appear broken. Men and women alike tinged the edges of their eyelids. And everyone smoked fat cheroots and chewed betel which turned their lips red and stained their teeth black.[7]

In contrast to later Western critics, Symes found the Burmans remarkably civilized. He observed their benevolence towards the infirm and the poor as well as their filial piety, which was 'strictly observed'. Moreover, he noted that Burmans were unusually tolerant of other religions, races, and ethnic groups. In sharp contrast, however, Symes was shocked at the Burmans' treatment of women. It was not at all unusual for fathers to sell their own daughters into prostitution to repay debts or to 'loan' their wives to foreigners as a sign of hospitality. Moreover, a man could have as many concubines as he could afford. His wife had no say in the matter. No better than a slave, a concubine lived under the same roof as her master's wife and became her property when the husband died.[8]

Like countless travellers before him, Symes journeyed to the ancient Shwe-dagon Pagoda, located about 2½ miles north of Rangoon (Colour Plate 14). The shrine's origins were inextricably linked to Burma's early conversion to Buddhism. According to legend, the pagoda was built by two merchants from southern Burma who offered Prince Gautama honey-cake following his enlightenment under the Bodhi tree. In return, the grateful Buddha gave the brothers eight hairs from his head with instructions to bury them near Thein-Got-Thara Hill not far from Rangoon. The brothers returned home and faithfully built a pagoda on the prescribed site to house the sacred relics. Then they set out to spread the Buddha's message among their countrymen. Over time the Shwe-dagon grew in size as Burma's kings renovated and enlarged it. By the middle of the fifteenth century, the pagoda had grown from its original height of 66 feet (20 metres) to 295 feet (90 metres). A devastating earthquake in 1779 severely damaged the shrine and King Hsinbyushin rebuilt it to a height of 326 feet (approximately 100 metres).[9] Royal artisans in Ava crafted a more elaborate golden tee (umbrella) mounted on a jewel-encrusted spire and, in grand procession, carried it to Rangoon accompanied by the king, his court, and a legion of monks. To make certain that the angry gods were appeased, the Burmans executed the former king of Pegu who had been held prisoner for over twenty years.[10]

This was the shrine that Symes visited in 1795. The road leading out of Rangoon to the Shwe-dagon was swept and in good repair. Along the way there were many charming small shrines erected by pilgrims. Several large, open monasteries sat back from the busy roadway shaded by pipal and tamarind trees. Symes's guide estimated that about 1,500 monks lived in the vicinity of the Shwe-dagon.[11]

Smaller temples, too numerous to count, surrounded the base of the larger pagoda, some quite new, some in ruins, and still others under construction. Each represented the effort of its builder to earn merit in the next world. From the pagoda's terrace, Symes had a wonderful view of the Pegu and Rangoon rivers winding through the 'level woody country'. In the distance, he could see several lakes. The pastoral scene was lovely. However, Symes left no doubt in the mind of his reader that he much preferred cultivated fields, the mark of an industrious and civilized society.[12]

The Anglo-Burmese Wars

Eighteenth-century Western visitors found the Burmans hospitable, friendly, and tolerant but a series of humiliating defeats by the British in the nineteenth century changed that. Imperial designs on Burma became apparent in the First Anglo-Burmese War (1824–6). Disputes along the Arakan frontier separating British India from Burma sparked hostilities in March 1824, but military activity quickly shifted to Lower Burma. Shortly after war broke out, a British expeditionary force moved up the Irrawaddy River and quickly occupied poorly defended Rangoon (Colour Plate 15). The flotilla would have advanced northward except for the onset of the monsoon season in May when things started going badly for the British. Widespread malaria, dysentery, and a severe shortage of supplies resulted in a high mortality rate and endless suffering for those who survived. The situation deteriorated until only several hundred out of 11,000 men were fit for duty.[13]

During the assault on Rangoon, British troops occupied the sacred Shwe-dagon Pagoda and fierce fighting ensued, resulting in heavy damage to the pagoda and surrounding structures. Casualties ran high, particularly for the ill-equipped and poorly trained Burmese. In April 1825, the British captured the city of Prome, and when the monsoon season ended, they rapidly advanced northward towards Amarapura. In order to save the capital, King Bagyidaw hastily agreed to sign the Treaty of Yandabo in February 1826. Burma's territorial losses to Britain were considerable, including Arakan, Tenasserim, Assam, and Manipur. The treaty also provided for a British Resident in Rangoon to oversee and protect the empire's trading interests.[14]

Following the war, British merchants found life in Rangoon frustrating and often dangerous. In December 1825, a fire had destroyed much of the town so that housing and shop space were in short supply. Westerners needed permission from the provincial governor, the Maywoon, to remain in the town, and officials seemed to delight in harassing them with new bureaucratic rules and regulations. Taxes were high and the usually good-natured natives treated the Westerners with contempt.[15]

By the late 1820s, Rangoon's population had fallen to 20,000, and the town was nothing more than a 'vast assemblage of wooden huts' surrounded by a high stockade and a ditch. The fort, if it could be called that, covered about 75 acres and had two principal streets, one where the Viceroy's run-down palace was located and Kaladan Street where foreigners kept their shops. All in all, there was little to recommend the town. Stagnant water pooled below the houses, giving off 'pestilential vapours', while 'herds of starving swine scavaged by day and packs of mean dogs' roamed the streets at night. The customs house, which Symes had found so impressive twenty-five years earlier, was 'fast tottering into ruins'. Beyond Rangoon, there was nothing but marshland and groves of jackfruit and pineapple. Puzun-daung, a small village of boatmen, lay a short distance to the east on the banks of the river by the same name.[16]

Despite official nagging and the lack of physical amenities, the British doggedly remained, their attention fixed on southern Burma's prospects for trade, her excellent ports, and her wealth of untapped interior resources. The avaricious merchants also deluded themselves by making plans for an overland trail to China.[17]

The Second Anglo-Burmese War began in 1852. Again fighting centred on the south. This time the British were well organized; supplies arrived on time and the troops remained healthy. The *Illustrated London News*, one of England's leading dailies, made no secret of Britian's designs on Burma, the next 'meal' for the Empire's voracious appetite. The 'populous and semi-savage kingdom of Burmah', the paper reported, 'promises to be the next instance of our capacity if not of our stomach for annexation'.[18]

Even before the war ended, the *Illustrated London News* began educating its readers about the soon-to-be colony. Rangoon and southern Burma promised to be valuable additions to the Empire. The natives would not be troublesome, the tabloid promised, for many had already expresed their desire to live under British rule.[19]

In often bellicose language, the newspaper described the furious battle for Rangoon which began on Easter Sunday in 1852. The redcoats performed brilliantly, leaving the native soldiers no other option than to flee or be taken prisoner. As they fled, the Burmese destroyed everything in their path. That mattered little to the British who held the prize, the Shwe-dagon Pagoda. Illustrations showed the victorious British troops, in their red wool uniforms and hats decorated with gilt, weapons in hand, standing about the pagoda.[20]

Confidently, the British captured the ports of Rangoon, Martaban, and Bassein, sweeping aside the Burmese army until they controlled the whole of southern Burma. On 20 December 1852, Major (later Commissioner) Sir Arthur Purves Phayre read a proclamation in Rangoon declaring the annexation of the Province of Pegu by the British government in Calcutta.[21]

Lower Burma became part of British India, and Lord Dalhousie, Governor-General of India, took great interest in the plans to refashion Rangoon into a colonial capital. The town was in near ruin and the few remaining buildings had little architectural value to the British. Consequently, Dalhousie could start with a 'clean sheet' to make Rangoon into 'one of the most beautiful cities and stations within the whole bounds of India'.[22]

Generally speaking, mid-nineteenth century British town planners sought to achieve certain well-recognized objectives by regulating the city's physical character. Goals included the promotion of commerce and communication, health and welfare, public safety, and perhaps most important, the civility of the inhabitants. Above all, order was the hallmark of the Victorian city. Streets were carefully laid out to intersect at right angles. Wharves, bridges, and commercial buildings were located properly adjacent to each other in order to smooth the flow of commerce. A broad strand along the river bank designed to prevent flooding was also a pleasant walkway; elsewhere in the city, parks and small gardens offered harried urbanites a welcome respite from the daily demands of city life. The result was that, despite differences in terrain and climate, from London to Calcutta the cities which made up the Victorian Empire's far-flung network of colonial seaports looked very much alike. The laying out of colonial Rangoon was not an experiment in town planning but a well-rehearsed process in which the symbol of the British Empire became neatly and irrevocably stamped on the map of Burma.

With myopic energy, the British turned Rangoon into what the *Illustrated London News* described as a 'British Burman Town'. Prior to victory in December 1852, British engineers were busy preparing plans for Rangoon. The immediate problem, first addressed by King Alaungpaya, was frequent flooding in the low-lying town. The British solution, found successful in London and other colonial cities, was the creation of a strand, a walkway 16 feet wide along the river bank. Between the strand and the river was a broad grassy area filled with palms and flowering shrubs to absorb the water whenever the river overflowed its banks. A network of drains also helped to forestall flooding.[23]

Under the direction of Lieutenant A. Fraser of the Bengal Engineers, Rangoon's old roads were repaired and new 'thoroughfares' constructed to improve the flow of traffic (Figure 7). With an eye for symmetry and beauty, Fraser widened the road encircling the old octagonal Shwe-sule Pagoda (Plate 15) and built roads radiating from it in four directions to make the pagoda the centre of Rangoon. Fraser designed the city's streets with enclosed conduits in the centre to drain off the rainwater. Early maps show Rangoon's streets intersecting neatly at right angles lined by uniform rectangular lots measuring 40 × 150 feet. The new government laid claim to most of the city's land and set about writing regulations for its redistribution and taxation.[24]

Preoccupied with rebuilding Rangoon, the British largely ignored the outwardly docile natives and demonstrated little interest in the habits and traditions of the people. The *Illustrated London News* summarily dismissed Burmese music as 'very monotonous', dancing as 'very peculiar', and the language as 'most strange . . . and most discordant'.[25]

One by one, the many urban institutions associated with Western cities appeared in Rangoon. In 1856, a municipal park, named for the city's benefactor, Lord Dalhousie, was laid out and the first government-sponsored school opened in a thatched shed. Within a year of the war's end, the Rangoon *Chronicle* began publication. Soon a group of merchants and civil servants established a

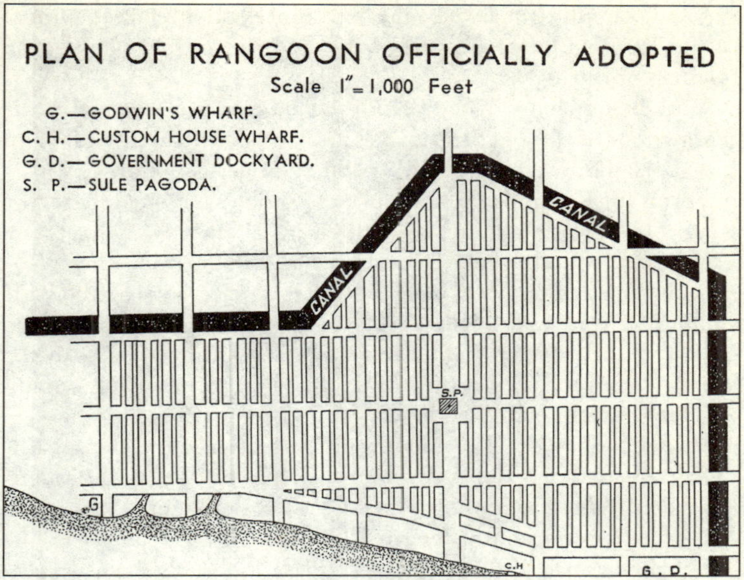

Fig. 7 Official plan of Rangoon. (Pearn, *History of Rangoon*, 1939)

Young Men's Literary Society which boasted a lending library for its members. In 1858, the British community enjoyed the first season of horse racing at the new turf club near Dalhousie Park.[26]

The *Illustrated London News* entertained readers with an account of an elephant steeplechase held in 1858 to celebrate Queen Victoria's birthday and the end of the officers' 'monsoon meeting' (Plate 16). Jockeys prodded the reluctant elephants into jumping across a broad ditch 7 feet wide. A sketch portrayed the merriment of the crowd as the lumbering and frightened animals tumbled headlong into the ditch.[27]

In the decades following annexation, the British borrowed heavily from native architecture to meet the urgent need for housing and office space. Simple raised houses of bamboo and wood with wide verandas open to the tropical breezes sprang up throughout Rangoon to accommodate the hundreds of civil servants and military officers who flooded into the city.

In 1863, the city's population was 61,000; within a decade it

15. The Shwe-sule Pagoda and Fytche Square, Rangoon. (Singer, *Burmah*, 1993)

16. An elephant steeplechase in Rangoon; engraving from the *Illustrated London News*, 25 September 1858. (Antiques of the Orient)

swelled to 100,000. Rangoon flourished despite frequent flooding, a steamy tropical environment, and a variety of life-threatening diseases. Commerce was clearly the main attraction (Colour Plate 16). Exports included teak, cotton, salt, and rubies. Not satisfied with an economy based solely on trade, enterprising merchants transformed the marshy delta to the south of Rangoon into rice paddies. It would not be long before rice would become the city's chief export. Just as vital to Rangoon's future, the British thought, was the need to convince King Mindon to allow them to pass through Upper Burma to reach China's huge markets via the Middle Kingdom's 'back door'.

By 1870, the so-called dacoits, roving bands of Burmese guerrillas who terrorized the countryside, were finally pacified and the British government was free to begin construction of a railroad system to facilitate trade and travel in Lower Burma. In Rangoon, Commissioner Phayre appointed a Municipal Committee composed of leading citizens and charged it with levying taxes, maintaining the roads, and managing the markets and hospitals. The lack of sewers and a safe water supply continued to trouble municipal officials.[28]

In 1870, a Lunatic Asylum was established and three years later the Rangoon General Hospital opened. Significantly, both the asylum and the hospital as well as other new municipal buildings were Victorian in design. The heavy buildings, with their ornamented façades, seemed more suitable to Birmingham and London and oddly out of place in tropical Rangoon. Throughout the British Empire, the ponderous Victorian-style architecture served to symbolize the oppressive permanence of British authority in the lives of the native people.

The imposition of British authority was not totally without benefits for the Burmans. In 1871, Phayre established a museum in Rangoon (Plate 17). Certainly, he hoped to place Rangoon on a par with other Victorian cities which all had museums. More than that, however, Phayre had a genuine scholarly interest in Burmese culture. He spoke the language fluently and was fascinated by the history, literature, and religion of the native people. The museum was located in a large native-style building in the centre of Rangoon and housed a wide variety of Burmese artefacts, includ-

17. Phayre Museum, Rangoon; engraving from the *Illustrated London News*, 4 May 1872. (Antiques of the Orient)

ing Buddhas from different time periods, native arts and crafts, and several royal thrones found in Lower Burma. Phayre further promoted interest in Burma among Western scholars in 1883 with the publication of a history of Burma, described by a contemporary as 'a remarkable piece of pioneer work'.[29]

Under British tutelage, Rangoon became a prosperous seaport, a brilliant jewel in the imperial crown. Nevertheless, the vast majority of the city's 130,000 inhabitants lived in squalor in mean alleyways and narrow side-streets. Crime, grinding poverty, and recurrent ethnic riots characterized life for most citizens in Lower Burma.

Motivated by the ever-present threat of urban violence, the authorities in Calcutta introduced self-government in Rangoon in 1882. Under the plan, Rangooners elected a Municipal Committee composed of representatives from the city's five major groups: Burmans, Europeans, Chinese, Muslims, and Hindus. However, the British were disappointed when most of the city's inhabitants did not vote, a failure which they attributed to apathy.[30] The real problem was that the British failed to recognize the negative impact that Burma's heavy-handed monarchs had had on the development of the people's natural political instincts.

Consequently and unfortunately, the British never attempted to educate the traditionally passive Burmans in the rights and responsibilities required to sustain and nurture democracy and republicanism. It was a failure which would have significant consequences for both the British and for Burma in the twentieth century.

Notwithstanding the political situation between the colonial government and the natives in Rangoon, relations between Lower Burma and Mandalay were cordial, a result no doubt of King Mindon's patience and perseverance. The diplomatic situation rapidly deteriorated, however, when the troubled and troublesome King Thibaw ascended to the throne of Upper Burma in 1878. Unwisely, the new king attempted to alter the diplomatic equation which divided the surrounding region, however unevenly, between British and French. Worse, he threatened Britain's monopoly on trade in Upper Burma. In 1885, the two nations went to war for the last time in the Third Anglo-Burmese War. The superior British forces easily captured Mandalay and sent King Thibaw and his wife into exile in India. Upper and Lower Burma were joined together in a single province within British India and Rangoon became the capital of Burma.[31]

The annexation of the north was measured not in casualties but in the significant growth in trade and shipping in the new capital city of Rangoon. Opportunities in commerce, shipping, and banking attracted growing numbers of immigrants and Rangoon's population increased to 248,000 by the turn of the century. It was a cosmopolitan city of Indians, Chinese, Afghans, Arabs, and Westerners who did business and occasionally clashed in vicious street riots when the 'ancient forces' of ethnicity got the better of them. One of the worst episodes on record occurred in 1893 when riots left thirty dead and some two hundred wounded. At the height of the fury, wrote one journalist, the steps of the mosque on Mogul Street (Plate 18) 'were slippery with the blood of mullahs and muezzins and chulias pouring out of ragged wounds'.[32] The rioters defeated the police and it became necessary to call in a regiment of British troops to stop the violence.

The Chinese occupied the neighbourhood of Latha and Maung Khaing Streets in what was described as 'a community of exclusive people, with an atmosphere and an architecture of its own; a com-

18. The mosque in Mogul Street, Rangoon, *c.*1890. (Singer, *Burmah*, 1993)

munity of rich merchants with broad views and the feelings of gentlemen'.[33] In the social taxonomy of Rangoon, the Chinese ranked immediately below the British who were said to prefer doing business with them rather than with the Indians or the Arabs. Ostentatiously, the Chinese imitated the British with their clubs and genteel behaviour. For that and their great wealth, they were admired and sometimes envied. Still, seasoned British merchants never let down their guard; they were certain that the Chinese could not be trusted.

Rangoon at the turn of the twentieth century was a study in modern urban development. It had been almost twenty years since steam-operated trams appeared. Telephone service as efficient as anywhere in the Empire, including London, was taken for granted. A modern hydropneumatic sewage system, completed in 1887, served that portion of the city frequented by the British, who generally turned a blind eye to the horrible sanitary conditions elsewhere in the capital. In 1904, safe drinking water became available with the opening of the Hlawaga Water Works, and three years later the business district was electrified. The litany of up-to-date city services continued with the construction of a new municipal hospital in 1911 to replace the old wooden structure, which was 'saturated by generations of microbes'. Two years later, motorized buses appeared on the streets of Rangoon.[34]

South of the city, land was reclaimed at great expense. New streets were laid out, and blocks of warehouses and shops were built as quickly as the land settled. At Monkey Point, where the Puzundaung Creek entered the Rangoon River, giant rice mills 'palpitated with life' every year from January to May, frantically processing nearly 2 million tons of rice from the delta paddies.[35]

Colonial Rangoon seemed to belong to everyone except the Burmans themselves. It was, in fact, a city without a history, conveniently without links to the country of which it was a part. Celebrating its modernity, one British resident wrote: 'It has no history to speak of; no buried past. Here is no "rose-red city, half as old as time"; but it is full of life and colour, a kaleidoscope of races with a growing character of its own, and the joyous atmosphere of youth.'[36]

Rangoon's worldly orientation had little attraction for the

natives who were overwhelmingly devout Buddhists. Their alienation was complete with the opening of the Victoria Memorial Park and Zoological Gardens just after the New Year in 1906. To Westerners, the zoo was a sure sign that colonial Rangoon had indeed come of age, for modern cities throughout North America and Europe alike either had or aspired to have a zoological garden. To the Buddhist Burmans, however, the very idea of caging wild animals for whatever purpose, educational or entertainment, was abhorrent.[37]

Increasingly, the alienated Burmans migrated to Kemendine north of the city along the banks of the Rangoon River. A contemporary account described the rural setting where 'the sculptor of alabaster Gautamas plies his chisel, the umbrella-maker displays all the delicate feeling of the race for beautiful things in the manufacture of yellow and green transparencies of perfect design, the weaver weaves tape for binding palm-leaf manuscripts into texts from the sacred books, the lacquer artist paints and gilds his cabinets for monastery libraries'. In Kemendine the culture of native Burma survived.[38]

The British were also retreating from the city at the turn of the century although for different reasons. Seeking respite from the bustle and heat of crowded Rangoon, upper-class merchants and lawyers built palatial villas on the hills overlooking the Royal Lake. Near by was Dalhousie Park, renowned for its spacious lawns and varied species of tropical trees and plants. One would have been hard-pressed to find a more exclusive neighbourhood in any of the empire's colonial capitals.[39]

Life in Rangoon went on pretty much as usual during World War I. Inhabitants complained about increased food prices and expressed concern when influenza cases were reported. The most serious problem, a dock strike, had little if anything to do with the war.[40]

Prosperity characterized Rangoon in the 1920s as it did most of the world's cities. The price of rice, the linchpin in Burma's economy, was high, a benefit even to small landholders, many of whom were Burmans. With rice exports booming, Rangoon experienced a facelift and many of the city's banks, clubs, and government offices were rebuilt. A number of new mansions dotted the hillsides around Victoria Lake in the suburbs, another indication of Rangoon's affluence.[41]

In retrospect, the major development of the post-war period was the founding of Rangoon University in 1920. Earlier opportunities for higher education in Burma were limited. There was the Rangoon Government College, a department of the Anglo-Vernacular High School, which prepared students to take external degrees from Calcutta University. Rangoon also has a small Baptist College as well as five normal schools for teacher training and a 'number of technical schools for surveying, elementary engineering, forestry and midwifery'.[42]

What was new after World War I was Sir Harcourt Butler's idea that Burma ought to have her own university, not an institution derived from or dependent upon a foreign university. Not surprisingly, the Governor's proposal for a new university was surrounded by controversy. Conservatives questioned the very need for a university in provincial Burma and warned that it would turn out too many unemployable lawyers and arts graduates. Radicals argued that the issue ultimately concerned home rule for Burma and independence from British India. The debate was ended when businessmen came out in support of a university in Rangoon. The Burmah Oil Company endowed an Engineering College, and merchants, traders, and businessmen made large contributions to Butler's University Fund. When it was established, Rangoon University brought together the work of the Government and Baptist Colleges and, during the 1920s, established departments in medicine, engineering, and forestry. The university's importance in Burma's national history can hardly be overstated. It was students from Rangoon University who ultimately led Burma to independence.[43]

'The year 1930 was a grim one,' wrote a British civil servant, 'giving us a taste, albeit faint, of the horrors in store in the future.'[44] The author was not referring to the Depression but instead to ethnic violence which occurred with greater frequency in the 1930s. Shortly after the start of the New Year, Indian dock workers went on strike for higher wages. Impoverished Burmans eagerly took their places at miserably low wages, but work slowed to a crawl because of the natives' lack of skills. Reluctantly, traders gave in to the Indians' demands. Upon hearing the news of the settlement, mobs of Burmans spilled into the streets pent-up with decades of anger and armed with 'iron fencing rails, sticks, (and) crowbars', eager to do battle with the Indians:

> Within a few moments a peaceful busy scene was transformed to one of horror. Men were hacked down, battered to death or hounded into the swift-flowing Irrawaddy to drown. Mobs of Burmans, armed and menacing, emerged from slum areas to hunt the men from southern India. The tenement buildings and barracks where they lived were invaded; women and children were massacred.[45]

In the melee, the police were 'swept aside, powerless to cope with the situation'. The violence went on for three days before British troops were called out. Quickly, almost effortlessly, they put down the now spent riots.

May brought more racial violence and an attempted jailbreak at the Rangoon jail. A pitched battle between convicts, who had somehow managed to secure weapons, and guards lasted for two hours before order was restored. Wary Rangooners armed themselves and talked nervously about how to protect their homes and families.[46]

Next came an earthquake preceded by a 'premonitory subterranean roar like the approach through a tunnel of some gigantic train' immediately followed by a 'petrifying tremor of the earth'.[47] Many of the city's old buildings collapsed completely; still others were severely damaged. Native houses and colonial bungalows built of timber swayed dangerously but withstood the shock. In Rangoon, forty deaths were recorded. To the south, less than 50 miles away, Pegu was completely destroyed.

The autumn of 1930 brought more racial violence, this time between the Burmans and the Chinese. The trouble was soon forgotten as a new threat loomed over Westerners in British Burma. During Christmas week, a native rebellion, expected at the outset to be short-lived, broke out in the countryside. The leader was a charismatic Burman named Saya San, who established his camp in the heavily forested steep mountains separating the Irrawaddy and the Sittang rivers. Following the tradition of King Alaungpaya, Saya San proclaimed himself king and set out with his ragtag army to take Burma back from the foreign invaders. He swiftly gained support from peasants and upper classes alike who offered money and assistance to the populist leader. Servants in private homes, waiters in British clubs, messengers in government offices, vendors in the bazaar—all eavesdropped and passed along bits and pieces of information through an incredibly efficient network to Saya San's headquarters. The revolt spread and it became necessary to bring in troops from India. Matching the Burmans in atrocities, British troops decapitated enemy bodies and hauled the bloody heads in sacks into nearby towns where they lined up the grotesque souvenirs side-by-side on a table as a grim warning to the natives.[48]

The British finally defeated Saya San, hunted him down, and took him to Rangoon, heavily guarded in handcuffs and leg irons. He was put on trial and duly executed. Ironically, the guerrilla leader's trial brought a new and more effective leader, Dr Ba Maw, into the national spotlight. Articulate, intelligent, and well-educated, Ba Maw played to the gallery as he defended Saya San in a British court. Ostensibly on the side of the law, he nevertheless managed to communicate his support for the rebel cause. Once the trial was over, Saya San became just another native rebel to the British, perhaps more popular than others before him and certainly more difficult to defeat, but nothing more than that. Only in retrospect would Ba Maw's role in the growing nationalist movement become clear.[49]

By the mid-1930s, Rangoon had a population of 342,000 and was the third leading seaport in India. The city became self-governing in 1935, but it was a meaningless achievement given the fact that 90 per cent of the city's population could not or did not vote. Few Westerners, however, concerned themselves with Burma's political and social inequalities. Instead, visitors praised Rangoon's 'tall, substantial buildings' and European appearance. They wrote postcards, journals, and articles for newspapers back home describing the city's thronging markets and serene suburban estates. They visited the rice paddies and the prosperous mills lining the bank along Puzun-daung Creek. With few exceptions, Western visitors left Rangoon singing the city's praises.[50]

Anti-British sentiment, which quietly festered in the 1920s and early 1930s, erupted in 1936 when university students, led by Aung San, went on strike. The leaders were determined to achieve independence. In their haste, Aung San and his colleagues struck a deal with the Japanese. The promise of independence was simply a cruel hoax. The Japanese betrayed the young Burmans and occupied the country during World War II. Burma thus exchanged one foreign master for another, the second much more tyrannical and deadly than the first. The experience convinced Aung San and his followers never again to accept colonial status for Burma. When the long struggle was over, a coalition of political parties headed by Aung San prepared to form a new government. On 4 January 1948, Burma became a sovereign independent republic.

Like Mandalay and other cities throughout South-East Asia, Rangoon suffered extensive damage during the war. Rangoon became the capital of newly independent Burma but it was a mere shadow of its old self. The new government had many problems and few resources and Rangoon deteriorated even further into a ghostly image of its imperial past. Crumbling colonial façades stood as testimony to a once great city.

It was a past, however, which made little sense to most Burmans. Rangoon built by and for the British embodied Western values and traditions. The bustling seaport bespoke Western economic theories, not harmony and serenity so important to Buddhists. The British may have built the city but they failed to ensure its longevity. Had they taught the native people even the most rudimentary facts about democracy and self-government, the outcome might have been different. But that is Rangoon's past, its phase as a colonial capital. Today Rangoon is once more called Yangon and the country is not Burma but Myanmar. A new spirit of nationalism seems to have taken hold in the land. It remains to be seen how the city will grow and change both physically and spiritually in the next few decades. One thing, however, is certain. The Yangon which emerges from this period of history will be a thoroughly Burmese city.

1. Michael Symes, *An Account of the Embassy to the Kingdom of Ava, 1795*, London: W. Bulmer and Co., 1800, p. 52.

2. Symes's mission had two goals. First, he settled problems along the frontier in Arakan and, second, he attempted to persuade the Court at Ava to close its ports

to French vessels. See D. G. E. Hall, *A History of South-East Asia*, 4th edn., London: Macmillan Education Ltd., 1981, pp. 628–30.

3. Symes, *Embassy to Ava*, pp. 456–62.

4. Ibid., pp. 183–5, 204–6; Hall, *History of South-East Asia*, pp. 630–2.

5. Symes, *Embassy to Ava*, pp. 204–5.

6. Ibid., pp. 185, 205.

7. Ibid., pp. 311–12, 328; Philip Shenon, 'Burma Cracks Down on a 2,000-year-old Beloved If Messy Vice', *International Herald Tribune*, 15 June 1995, pp. 1, 10.

8. Symes, *Embassy to Ava*, pp. 216–18, 313, 318.

9. Wilhelm Klein et al., *Burma*, Hong Kong: APA Publications, Insight Guides, 1988, pp. 125–6.

10. Symes, *Embassy to Ava*, pp. 82–3.

11. Ibid., p. 209.

12. Ibid., pp. 207–9. Symes observed that more merit was earned by building a new temple than by repairing an old one. This explained the large number of tumbledown or dilapidated shrines while new ones were under construction near by.

13. Hall, *History of South-East Asia*, pp. 637–41.

14. Ibid.; J. J. Snodgrass, *Narrative of the Burmese War*, London: John Murray, 1827, p. 121.

15. Snodgrass, *Burmese War*, p. 13.

16. Ibid., pp. 12–14, 205.

17. Ibid., pp. 286–91.

18. *Illustrated London News*, 17 April 1852; Hall, *History of South-East Asia*, p. 642. Hall denies that the British had any designs on Burma after the first war and states that they made every effort to maintain peaceful relations. He adds, however: 'What they failed to realize was that once they had a foothold in the country the sheer force of circumstances was bound ultimately to bring about complete annexation no matter how unwilling they were to extend their territorial commitments.'

19. *Illustrated London News*, 21 August 1852.

20. Ibid., 26 June 1852.

21. Hall, *History of South-East Asia*, pp. 650–1.

22. Ibid., p. 658; Bertie Reginald Pearn, *A History of Rangoon*, 1939; reprint edn., London: Farnborough Gregg, 1971, pp. 176–80.

23. Pearn, *History of Rangoon*, p. 183; *Illustrated London News*, 4 December 1852.

24. Pearn, *History of Rangoon*, pp. 183–4; HRH Prince Damrong Rajanubhab, *Journey through Burma in 1936*, trans. Kennon Breazeale, Bangkok: River Books, 1991, pp. 30–1.

25. *Illustrated London News*, 5 June 1852 and 7 May 1853.

26. Pearn, *History of Rangoon*, pp. 184–5.

27. *Illustrated London News*, 25 September 1858.

28. Pearn, *History of Rangoon*, pp. 228–9.

29. Hall, *History of South-East Asia*, p. 655: Damrong, *Journey through Burma*, p. 28.

30. Pearn, *History of Rangoon*, pp. 255–6.

31. Hall, *History of South-East Asia*, pp. 679–82.

32. V. C. Scott O'Connor, *The Silken East: A Record of Life and Travel in Burma*, 2 vols., London: Hutchinson and Co., 1904, Vol. 1, pp. 69–72.

33. Ibid., pp. 71–2.

34. Pearn, *History of Rangoon*, pp. 271–8.

35. O'Connor, *The Silken East*, pp. 91–2; Hall, *History of South-East Asia*, pp. 821–4.

36. O'Connor, *The Silken East*, p. 76.

37. Pearn, *History of Rangoon*, p. 278.

38. O'Connor, *The Silken East*, pp. 72–4.

39. Ibid., pp. 76–7.

40. Hall has very little to say about the effects of World War I on Burma.

41. E. C. V. Foucar, *I Lived in Burma*, London: Dennis Dobson. 1956, p. 40.

42. Hall, *History of South-East Asia*, p. 776.

43. Foucar, *I Lived in Burma*, p. 41; Hall, *History of South-East Asia*, pp. 776–7.

44. Foucar, *I Lived in Burma*, p. 65.

45. Ibid., p. 66.

46. Ibid., pp. 67–8.

47. Ibid., pp. 58–9.

48. Ibid., pp. 69–75.

49. Ibid., p. 75.

50. Damrong, *Journey through Burma*, pp. 29–30; Pearn, *History of Rangoon*, p. 295; Hall, *History of South-East Asia*, pp. 784–5.

Bibliography

Aung San Suu Kyi, *Freedom from Fear and Other Writings*, ed. Michael Aris, Harmondsworth: Penguin Books, 1991.

Aung-Thwin, Michael, 'Hierarchy and Order in Pre-colonial Burma', *Journal of Southeast Asian Studies*, XV (Sept. 1984): 224–32.

———, *Historical Sites in Burma*, Rangoon: Ministry of Union Culture, Government of the Union of Burma, 1972.

———, 'Kingship, the Sangha, and Society in Pagan', in Kenneth R. Hall and John K. Whitmore (eds.), *Explorations in Early Southeast Asian History: The Origins of Southeast Asian Statecraft*, Ann Arbor: Center for South and Southeast Asian Studies, University of Michigan, 1976.

———, *Pagan: The Origins of Modern Burma*, Honolulu: University of Hawaii Press, 1985.

Coedès, George, *Angkor: An Introduction*, trans. Emily Floyd Gardiner, 1961; reprint edn., Singapore: Oxford University Press, 1986.

Conze, Edward, *A Short History of Buddhism*, London: George Allen and Unwin, 1980.

Damrong, HRH Prince Rajanubhab, *Journey through Burma in 1936*, trans. Kennon Breazeale, Bangkok: River Books, 1991.

Fielding, H., *Thibaw's Queen*, London: Harper and Brothers, 1899.

Foucar, E. C. V., *I Lived in Burma*, London: Dennis Dobson, 1956.

———, *Mandalay the Golden*, London: Dennis Dobson, 1963.

Fytche, Albert, *Burma Past and Present*, 2 vols., London: C. Kegan Paul and Co., 1878.

Hall, D. G. E., *A History of South-East Asia*, 4th edn., London: Macmillan Education Ltd., 1981.

Illustrated London News, London, 1842–1900.

Keeton, Charles Lee, III, *King Thibaw and the Ecological Rape of Burma*, Delhi: Manohar Book Service, 1974.

Kelly, R. Talbot, *Burma*, 1905; 2nd edn., London: A & C Black, Ltd., 1933.

Klein, William et al., *Burma*, Hong Kong: APA Publications, Insight Guides, 1988.

Lehman, F. K. (Chit Hlaing), 'Freedom and Bondage in Traditional Burma and Thailand', *Journal of Southeast Asian Studies*, XV (Sept. 1984): 233–44.

Luce, G. H., 'The Greater Temples of Pagan', in *Fiftieth Anniversary Publications*, No. 2, Rangoon Burma Research Society, 1960.

———, *Phases of Pre-Pagan Burma: Languages and History*, 2 vols., London: Oxford University Press, 1985.

Moore, Lt. Joseph and Marryat, Cpt. Frederick, *Birman Empire, Parts I–III, and Combined Operations of the British Forces in the Birman Empire, 1824 and 1825*, London: Kingsbury and Co., 1825.

O'Connor, V. C. Scott, *Mandalay and Other Cities of the Past in Burma*, Hutchinson and Co., London: 1907; reprint edn., Bangkok: White Lotus Co., 1987.

———, *The Silken East: A Record of Life and Travel in Burma*, 2 vols., London: Hutchinson and Co., 1904.

Pe Maung Tin and Luce, G. H. (trans.), *The Glass Palace Chronicle of the Kings of Burma*, London: Oxford Unviersity Press, 1923.

Phayre, Arthur P., *History of Burma: Including Burma Proper, Pegu, Taungu, Tenasserim, and Arakan*, 1883: reprint edn., New York: Augustus M. Kelley Publishers, 1969.

Pichard, Pierre, *The Pentagonal Monuments of Pagan*, Bangkok: White Lotus Co., 1991.

Pearn, Bertie Reginald, *A History of Rangoon*, 1939; reprint edn., London: Farnborough Gregg, 1971.

Scott, J. G., *Burma from the Earliest Times to the Present Day*, London: T. Fisher Unwin Ltd., 1924.

Seagrave, Sterling, *Dragon Lady: The Life and Legend of the Last Empress of China*, New York: Vintage Books, 1993.

Shenon Philip, 'Burma Cracks Down on a 2,000-year-old Beloved If Messy Vice', *International Herald Tribune*, 15 June 1995.

Shway Yoe [Sir George Scott], *The Burman: His Life and Notions*, 1882; reprint edn., Gartmore: Kiscadale Publications, 1989.

Singer, Noel F., *Burmah: A Photographic Journey, 1855–1925*, Gartmore: Kiscadale Publications, 1993.

Snodgrass, J. J., *Narrative of the Burmese War*, London: John Murray, 1827.

Strachan, Paul, *Mandalay: Travels from the Golden City*, Gartmore: Kiscadale Publications, 1994.

———, *Pagan: Art and Architecture of Old Burma*, Gartmore: Kiscadale Publications, 1989.

Symes, Michael, *An Account of the Embassy to the Kingdom of Ava, 1795*, London: W. Bulmer and Co., 1800.

Tarling, Nicholas, *A Concise History of Southeast Asia*, Singapore: Donald Moore Press Ltd., 1969, pp. 3–8.

Vincent, Frank, *The Land of the White Elephant: Sights and Scenes in South-Eastern Asia*, New York: Harper & Brothers, 1874.

Yule, Henry, *A Narrative of the Mission to the Court of Ava in 1855*, Calcutta: Thacker, Spink, and Co., 1857.

Index

Other Oxford Paperbacks for readers interested in South-East Asia, past and present

Cambodia

Angkor: An Introduction
GEORGE COEDÈS

Angkor and the Khmers
MALCOLM MacDONALD

Indonesia

An Artist in Java
JAN POORTENAAR

Bali and Angkor
GEOFFREY GORER

Coolie
MADELON H. LULOFS

Diverse Lives
JEANETTE LINGARD

Flowering Lotus
HAROLD FORSTER

Forever a Stranger and Other Stories
HELLA S. HAASSE

Forgotten Kingdoms in Sumatra
F. M. SCHNITGER

The Head-Hunters of Borneo
CARL BOCK

The Hidden Force*
LOUIS COUPERUS

The Hunt for the Heart
VINCENT MAHIEU

In Borneo Jungles
WILLIAM O. KHRON

Island of Bali*
MIGUEL COVARRUBIAS

Java: Facts and Fancies
AUGUSTA DE WIT

Java: The Garden of the East
E. R. SCIDMORE

Java: A Travellers' Anthology
JAMES R. RUSH

The Last Paradise
HICKMAN POWELL

Let It Be
PAULA GOMES

Makassar Sailing
G. E. P. COLLINS

The Malay Archipelago
ALFRED RUSSEL WALLACE

The Outlaw and Other Stories
MOCHTAR LUBIS

The Poison Tree*
E. M. BEEKMAN (Ed.)

Rambles in Java and the Straits in 1852
'BENGAL CIVILIAN' (C. W. KINLOCH)

Rubber
MADELON H. LULOFS

A Tale from Bali*
VICKI BAUM

The Temples of Java
JACQUES DUMARÇAY

Through Central Borneo
CARL LUMHOLTZ

To the Spice Islands and Beyond
GEORGE MILLER

Travelling to Bali
ADRIAN VICKERS

Twin Flower: A Story of Bali
G. E. P. COLLINS

Unbeaten Tracks in Islands of the Far East
ANNA FORBES

Witnesses to Sumatra
ANTHONY REID

Yogyakarta
MICHAEL SMITHIES

Malaysia

Among Primitive Peoples in Borneo
IVOR H. N. EVANS

An Analysis of Malay Magic
K. M. ENDICOTT

At the Court of Pelesu
HUGH CLIFFORD

The Best of Borneo Travel
VICTOR T. KING

Borneo Jungle
TOM HARRISSON

The Chersonese with the Gliding Off
EMILY INNES

The Experiences of a Hunter
WILLIAM T. HORNADAY

The Field-Book of a Jungle-Wallah
CHARLES HOSE

50 Years of Romance and Research
CHARLES HOSE

The Gardens of the Sun
F. W. BURBIDGE

Glimpses into Life in Malayan Lands
JOHN TURNBULL THOMSON

The Golden Chersonese
ISABELLA BIRD

Illustrated Guide to the Federated Malay States (1923)
C. W. HARISSON

The Malay Magician
RICHARD WINSTEDT

Malay Poisons and Charm Cures
JOHN D. GIMLETTE

My Life in Sarawak
MARGARET BROOKE,
THE RANEE OF SARAWAK

Natural Man
CHARLES HOSE

Nine Dayak Nights
W. R. GEDDES

A Nocturne and Other Malayan Stories and Sketches
FRANK SWETTENHAM

Orang-Utan
BARBARA HARRISSON

The Pirate Wind
OWEN RUTTER

Queen of the Head-Hunters
SYLVIA, LADY BROOKE,
THE RANEE OF SARAWAK

Six Years in the Malay Jungle
CARVETH WELLS

The Soul of Malaya
HENRI FAUCONNIER

They Came to Malaya
J. M. GULLICK

Wanderings in the Great Forests of Borneo
ODOARDO BECCARI

The White Rajahs of Sarawak
ROBERT PAYN

Myanmar

Faded Splendour, Golden Past: Urban Images of Burma
ELLEN CORWIN CANGI

Philippines

Little Brown Brother
LEON WOLFF

Singapore

Manners and Customs of the Chinese
J. D. VAUGHAN

Raffles of the Eastern Isles
C. E. WURTZBURG

Singapore 1941–1942
MASANOBU TSUJI

Travellers' Singapore
JOHN BASTIN

South-East Asia

Adventures and Encounters
J. M. GULLICK

Adventurous Women
J. M. GULLICK (Ed.)

Explorers of South-East Asia
VICTOR T. KING (Ed.)

Soul of the Tiger*
J. A. McNEELY and P. S. WACHTEL

Wonders of Nature in South-East Asia
THE EARL OF CRANBROOK

Thailand

Behind the Painting and Other Stories
SIBURAPHA

Descriptions of Old Siam
MICHAEL SMITHIES

The English Governess at the Siamese Court
ANNA LEONOWENS

The Politician and Other Stories
KHAMSING SRINAWK

The Prostitute
K. SURANGKHANANG

Temples and Elephants
CARL BOCK

To Siam and Malaya in the Duke of Sutherland's Yacht *Sans Peur*
FLORENCE CADDY

Travels in Siam, Cambodia and Laos
HENRI MOUHOT

Vietnam

The General Retires and Other Stories
NGUYEN HUY THIEP

The Light of the Capital
GREG & MONIQUE LOCKHART

**Titles marked with an asterisk have restricted rights.*